OBSERVATIONAL MANIFESTATION OF CHAOS
IN ASTROPHYSICAL OBJECTS

Invited talks for a workshop held in Moscow,
Sternberg Astronomical Institute,
28–29 August 2000

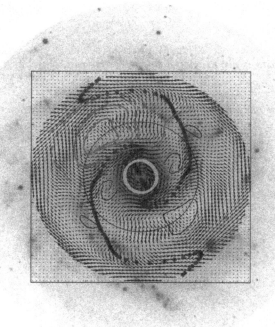

Cover illustration: R-band image of the spiral galaxy NGC 3631 as obtained from the
ING archive. Asterisks show the locations of the maxima of the second Fourier harmonic
of the R-band brightness map of the galaxy. The black thin circle marks the position of
the corotation. The black solid lines demonstrate the vortex separatrices or nearly closed
streamlines in the absence of a separatrix. The two-dimensional velocity field is marked
by arrows with some trajectories superimposed in the reference frame rotating with the
pattern speed. Gray solid lines display: 1) a closed separatrix around an anticyclone, 2) a
reference trajectory that starts from the separatrix. Gray dashed lines show some trajectories
that diverge exponentially from the reference trajectory. All lines belong to a family of the
chaotic trajectories. The spectrum of these stochastic trajectories is shown in the upper right
corner of the figure. The ring is a family of the regular trajectories.

OBSERVATIONAL MANIFESTATION OF CHAOS
IN ASTROPHYSICAL OBJECTS

Invited talks for a workshop held in Moscow, Sternberg Astronomical Institute,
28–29 August 2000

Edited by

Alexei M. Fridman
Institute of Astronomy, Russian Academy of Sciences, Moscow, Russia

Mikhail Ya. Marov
M.V. Keldysh Institute of Applied Mathematics, Russian Academy of Science, Moscow,
Russia

Richard H. Miller
Astronomy and Astrophysics Center, University of Chicago, Chicago, USA

Reprinted from *Space Science Reviews*, Volume 102, Nos. 1–4, 2002

Springer-Science+Business Media, B.V.

A C.I.P. Catalogue record for this book is available from the Library of Congress

ISBN 978-94-010-3945-1   ISBN 978-94-010-0247-9 (eBook)
DOI 10.1007/978-94-010-0247-9

Printed on acid-free paper

ISBN 978-94-010-3945-1   ISBN 978-94-010-0247-9 (eBook)
DOI 10.1007/978-94-010-0247-9

Printed on acid-free paper

SPACE SCIENCE REVIEWS / *Vol. 102   Nos. 1–4   2002*

TABLE OF CONTENTS

TABLE OF CONTENTS

Foreword

Foreword

On August 2000 in the Lomonosov Moscow State University the first scientific conference dedicated to chaos in the real astronomical systems was held. The most prominent astrophysisists – specialist in the field of stochastic dynamics – attended the conference. A broad scope of the problems related to the observed manifestations of chaotic motions in galactic and stellar objects, with the involvement of basic theory and numerical modeling, were addressed.

The idea (not so obvious, as we believe, to many astrophysicists) was to show that, while great progress in the field of stochastic mechanics was accomplished, the science of chaos in actually observed systems is only just being born.

Basically, the situation described prompted the organizers to hold the meeting in order to discuss chaotic processes in real systems.

It seemed worthwhile to begin these introductory remarks with a brief description of some events that preceeded the conference.

Since actually existing systems are the subject of the natural sciences, and in the latter experiments play the key role, we shall begin our account with the experimental results.

At the beginning of 1960s, in the Novosibirsk Institute of Nuclear Physics (where one of the authors, AMF, has worked for ten years), its former Director Academician Budker initiated the study of stability conditions of charged particle motions in a cyclic accelerator. The motion of a particle in the accelerator can be considered as vibrations of an oscillator with two degrees of freedom, undergoing some periodic perturbations resulting from imperfections in the accelerator's magnetic field.

The main question in the impending investigations was: what are the conditions for stable oscillations of the particles without their loss in the process of acceleration?

At that time theoretical studies of stability were carried out in the so-called linear approximation, i.e. assuming that the frequency of oscillations does not depend on their amplitude. This is true, for instance, for an ordinary pendulum when the amplitude of the oscillations is sufficiently small. In this particular case, the worst threat is resonance which emerges when the frequency of an external perturbation coincides with that of the oscillation. In application to the accelerator, such a situation results in the linear growth of the amplitude of oscillations with the time ($\sim t$), and gives rise to a loss of particles.

In general case, however, the oscillations are nonlinear: their frequency depends on the amplitude. The frequency of oscillations of a pendulum, for instance, decreases when the amplitude increases, and it tends to zero when the amplitude reaches its maximum (at 180°). At the first sight, it looks like an opportunity to overcome the 'resonance threat' since the frequency variation takes the oscillator away from the resonance and thus limits the growth of the amplitude of oscillations. However, the very first experimental investigations have already shown that in such a case a new, rather unusual (from the basic mechanics viewpoint) instability arises. Oscillations become irregular, their amplitude growing in average proportional to $\sqrt{t}$, and large fluctuations appear. It looks as if the oscillator was subjected to a random perturbation, while in reality the perturbation is strictly periodical in terms of time. Following Chirikov, this unusual process was referred to as a stochastic instability of the nonlinear oscillations. The further study of this instability showed that the motion of a fully determined mechanical system can be, to say the least, very much like a random process. This kind of motion only appears provided the mechanical system is unstable.

The above example had been initially thought to be the very first known mechanism indicating that statistical laws are applicable to a dynamical system. However, as it became clear later on, similar phenomena were found and described by some scientists much earlier. It seems that a relationship between the statistical laws and an instability of the mechanical motion was first noted by Poincaré. The same problem has been thoroughly investigated by the Russian scientist Krylov and though his early death in 1947 prevented him from completing this pioneering study, the results obtained are of extreme interest. In the following years, many authors focused on the problem, the main breakthroughs being made by Kolmogorov, Sinai, Anosov, Arnold, Smale, Moser, and some others.

Let us now define precisely what we mean by term 'stochastic motion'. This type of motion is characterized by the following features:

- *Ergodicity.* This means that the trajectory covers uniformly a certain area of the phase space, although in itself, this property poorly conforms to our idea of a random process. As an example, the electronic ray motion on a TV screen is ergodic but certainly not random.

- *Local instability.* This is the strongest type of instability, possible attributing even to the finite motion of a dynamical system or its parts. It is due to this instability that dynamical stochasticity, or chaos, arises. Let us denote by $D(t)$ the distance between two points occupying two different (and close to each other at the initial moment of time $t = 0$) trajectories in the phase space. The local instability manifests itself when the distance between these trajectories grows exponentially in terms of time:

$$D(t) = D(0)e^{\gamma t}. \tag{1}$$

The growth rate of instability γ is a function of the point in the phase space. The exponential divergence of the trajectories in a given system can emerge not for all initial conditions. However, a local instability means that there exists a domain of finite measure such that if we choose any point of this domain as the initial point, its small perturbation results in the great divergence of the corresponding trajectories.

- *Mixing of phase trajectories, KS (Kolmogorov – Sinai) entropy, and their re-lationship to each other as well as to the local instability.* The property to possess a local instability is directly related to the property of having mixed phase trajectories. Indeed, if the motion of a system is finite, the initially close and then exponentially diverging (due to local instability) trajectories can not get apart beyond the size of the area where the motion occurs. This confinement leads to strong intermixing of trajectories and dramatic confusion of the patterns. For instance, the area of phase 'drop' in the form of a sphere, having occupied an initial phase volume $\delta\Gamma_0$, within a short period of time becomes severely indented.

The degree of complication grows with time and can bve evaluated by the rate of mixing. In order to take into account the powerful complicaiton of the shape of a phase drop quantitatively let us introduce a notion of the envelope phase volume $\Delta\Gamma(t)$. Then, using the formula of the local instability, we can evaluate

$$\overline{\Delta\Gamma(t)} = \Delta\Gamma_0 e^{ht}. \tag{2}$$

Correspondingly, the entropy of a phase drop upon 'coarse-grained' averaging is:

$$S = \ln\overline{\Delta\Gamma(t)} = \ln(\Delta\Gamma_0 e^{ht}) = ht + \ln\Delta\Gamma_0. \tag{3}$$

The value $\Delta\Gamma_0$ can be put equal to small value of the volume over which the averaging is performed. Inside this volume, the 'bubbles of emptiness' turn out to be imperceptible. The expression

$$\lim_{\Delta\Gamma_0\to 0}\ \lim_{t\to\infty}\frac{1}{t}\ln(\overline{\Delta\Gamma(t)}) = \lim_{\Delta\Gamma_0\to 0}\ \lim_{t\to\infty}\frac{1}{t}(ht + \ln(\Delta\Gamma_0) = h \tag{4}$$

defines the Kolmogorov – Sinai (KS) entropy. The definition of this entropy was introduced in 1958–59 by Kolmogorov and, upon some correction, in 1959 by Sinai.

The mathematical expression (4) determines also the rate of the entrophy S change in the 'envelope' phase volume. As follows from (1), this rate coincides with the growth rate of the local instability. Therefore, the three characteristic time periods: mixing τ, local instability and change of the entropy S, turn out also in coincidence:

$$\tau \sim \frac{1}{\gamma} \sim \frac{1}{h}. \tag{5}$$

In stochastic dynamics the Lyapunov characteristic number (LCN), λ, plays an important role. According to the definition

$$\lambda = \lim_{D(0)\to 0}\ \lim_{t\to\infty}\frac{1}{t}\ln D(t)/D(0), \tag{6}$$

a positive value of the LCN for a family of the initialy nearby trajectories means they follow an exponential divergence, i.e. the existence of the local instability. By the order of magnitude, the growth rate of the local instability γ (and, correspondingly, the value of the KS-entrophy h), corresponds to the value of the Lyapunov number λ:

$$\gamma \sim h \sim \lambda. \tag{7}$$

The expression (5) and (7) relate the local instability, mixing and the KS entropy with LCN. They play a fundamental role in the analysis of conditions for chaos to be set up in a dynamical system.

Being an intrinsic property of the dynamical system, chaos can arise almost everywhere and every time. While it is not always detected in numerical experiments, this is caused by either unresolvably thin area of the parameters involved, or limited time of the observations. Sometimes it could be also obscured by other more powerful overlapping processes.

The area of applications of the phenomenon of stochasticity turned out to be unexpectedly wide. Having originated from statistical physics, it encompasses virtually all main branches of contemporary classical and quantum physics, including astrophysics, biophysics, chemical physics and so on.

The relationship between dynamics and stochasticity has been fascinating physicists as long for as a century. The main question to be answered was essentially the same: it is possible to rigorously derive the statistical description from the dynamical one? Until most recently the answer was no. With regard to the classic (non-quantum) dynamical system not subjected to noise, the appearance of a random factor was used to be assigned exclusively to its complexity. The latter was associated with an extremely big number of degrees of freedom (gas in a vessel could serve as an example), when the deterministic description becomes simply meaningless, though in principle still possible. In this case, transition to a probabilistic description was usually based upon some hypothesis, such as the ergodic one.

The rigorous theory that has now emerged allows us to claim that the nonlinear dynamical system can 'generate' statistics. In other words, the statistical approach provides an adequate description of the real behaviour of a dynamical system, rather than just serving as a method of its approximate analysis. The striking recent progress in physics and mathematics came with the understanding that even in very simple nonlinear systems (possessing a small number of the degrees of freedom) random behaviour may appear. Brilliant examples are a ball in billiards with concave walls (Snai billiards), an electron in the field of two sinusoidal electromagnetic waves, etc. How does a random factor emerge in a deterministic system, in spite of the fact that the solution is unique and single-valued? A short answer is that it is because of instability of the individual motions occuring in a limited phase volume. Instability leads to the complexity and diversity of almost all kinds of individual motions, so that the concept of ensemble naturally arises for which the statistical description is valid.

At present, the theory was developed, with its main points being supported by experiment. The theory describes transition from deterministic to stochastic behaviour for the most diverse systems, such as hydrodynamic flows, radio generators of random signals, autocatalytic chemical reactions, etc. It seems natural to suppose that if the behaviour of a system with small number of the degrees of freedom is complex, then a system with an unlimited number of the degrees of freedom would unavoidably demonstrate random behaviour. However, generally it is not the case.

Once there was a hypothesis put forward proposing that in systems with very large numbers of the degrees of freedom, a weak nonlinearity would be sufficient to distribute over all the modes the energy stored in the individual degrees of freedom and hence, to establish thermodynamic equilibrium. In order to support this idea, as early as at the end of 1940s a series of numerical experiments was carried out based on the model of nonlinear chains of large number of particles. However, thermalization was not found – the system periodically returned to the state with the initial distribution of energy (Fermi-Pasta-Ulam paradox). In reality, the nonlinear wave systems are of two types: the integrable (or close to such) systems which exhibit only simple periodical or quasi-periodical patterns, and nonintegrable systems. The latter become stochastic when a sufficiently large initial energy is supplied. Interestingly enough, the system which Fermi, Pasta and Ulam have focused on, involving the chain model and the values of the parameters selected, turned out to be close to an integrable one.

In both integrable and nonintegrable systems, particular solutions may exist which correspond to the so-called coherent formations or spatial structures (solitons, stationary shock waves, etc). Coherent nonlinear formations are now studied in detail in the solid bodies (domains), in plasma (Langmuir and Alfven solitons), in planetary athmospheres and galactic disks (cyclones and anticyclones), in nonlinear optics (ultrashort pulses). There is an idea viewing elementary particles as solitons of the quantum fields that will be hopefully confirmed. Of specific interest are coherent formations in dissipative nonequilibrium media – the dissipative structures and autowaves (the term introduced by Khokhlov by analogy with the autooscillations). Examples of such autowaves and dissipative structures are: waves of fire; pulses of excitation in the neural and muscular fibres; the change (in terms of space and time) of the number of species in the populations of living organisms; concentrational waves in autocatalytic chemical reactions. The main peculiarity of those time-space structures is that they do not much depend on the properties of the source of disturbance and initial state of the underlying medium, as well as on boundary conditions. Dissipative structures in nonequilibrium media can be regarded as one of the typical and natural forms of self-organization.

Returning now to the most essential question we deal with in these Proceedings – arising of chaos in real astrophysical systems – let us recall that it is the Lyapunov characteristic numbers that serve as the quantative measure of stochasticity in the system. As follows from (6), the definition of the Lyapunov numbers encompasses two conditions (limiting processes): tending to zero of the initial distance between two points chosen on two close trajectories; and tending to infinity of the time of divergence of the points on those trajectories. Obviously, for a real object with finite angular resolution and finite time of observation, the above two conditions can not be fulfilled. Therefore, the proofs of some well-known theorems on the properties of the Lyapunov characteristic numbers, in particular as the said passages to limits are concerned, have to be comprehended in a different context in the case of real systems. Indeed, the finiteness of the observational time requires utilization of the adequate computational algorithms. The recently discovered coherent structures (giant cyclones and anticyclones) in galactic disks and chaotic trajectories (exponentially diverging streamlines) co-exist in the vicinity of the separatrix. Because the latter separates the "trapped" liquid particles (in the coherent structures) from the "untrapped" ones (in the chaotic trajectories), the simultaneous study of the properties of the integrable and nonintegrable systems is suggested. Respectively, particularities of the solutions of both ordinary differential equations and partial differential equations are to be investigated, which entails great difficulties.

Let us now briefly review at least some results of the papers included in the Proceedings.

Of great interest are the results of investigation of the peculiarities of chaotic system behaviour in the process of numerical solution of the equations of motion with various accuracies, including evaluation of the time scale in the process of reaching an equilibrium.

Considerable progress is made in the study of the integrals of motion in self-consistent systems, in particular the various forms of invariant curves for both resonance and non-resonance systems, using truncated forms of the third integral.

Very important are the results of numerical modeling the process of evolution of the disk systems based on the nonlinear theory of resonance excitation of density waves. The effects of self-gravitation and viscosity were taken into account the boundary conditions of the problem being varied. In particular, it was shown that resonance excitation is a very effective mechanism, and even a weak perturbation exerts a strong influence upon the morphology of such nonlinear systems, the density waves themselves playing an important role in the disk's evolution.

The feasibility of observing chaotic behavior in the stellar component of spiral galaxies is discussed with account for both the modern theoretical expectations on the sources for chaos development and the current state-of-the-art observational equipment.

A galaxy that would otherwise be in equilibrium is shown to be particularly sensitive to external perturbations if an appreciable fraction of its stars are on chaotic orbits. This destroys the equilibrium and leads to slow drifts in the global dynamics of the galaxy.

Departures from quasi-periodic motions of the individual stars in a galaxy model provide a sensitive test to detect the presence of chaotic orbits. Discrete temporal Fourier transforms of a sample of stars speeds this analysis and improves its accuracy.

For the gaseous disk of the spiral galaxy NGC 3631 (taken as an example), and existence of the chaotic trajectories in the vicinity of the separatrix (separating the areas of the giant vortices, consisting of trapped gaseous "particles", from the rest area of untrapped "particles", in the configurative space), was proved on the basis of calculation of the characteristic Lyapunov numbers in the quasi-stationary field of the streamlines. The latter were reconstructed from the results of the observations of the line-of-sight velocity field. This study greatly contributes to our idea of the large-scale structure of the velocity field of a spiral galaxy, clarifying its dynamical portrait.

Observational data on stars and high-energy sources were represented by a review of specific formations demonstrating their stochastic nature, which agrees with the results of modeling. The peculiarities of the precessional motion of the accretional disk and relativistic jets in the double system SS 433 were analysed in detail. Some results of the spectral and photometric observations of the system can be attributed to stochastic processes.

Unfortunately, by some or other reasons, not all the results reported at the conference could be included in the volume. The topic of chaos manifestation in the real environment is very broad and continuously extends in scope though the processes involved not always clearly emerge. For example, it was of great interest some specific phenomena related to adiabatic piston moving under the action of elastic collisions with the particles of the gases, which was discussed by Ya.G. Sinai. While the problem of deriving the equation describing the dynamics of the piston was popular a hundred years ago, significant progress was achieved only in recent times when an integrable case was derived and the convergence to some non-reversible behaviour was proved.

We hope that the results of this volume will be met with an interest by the researches working in the different fields. For astrophysicists, they will serve as an incentive to study the stochastic nature of astronomical objects, whereas for mechanics they will open new frontiers for the nonlinear dynamics application.

We are grateful to all the authors for their cooperation in the process of preparation and feedback when editing the manuscripts, and to the reviewers whose interest and high professionalism promoted to the improvement of the papers quality. We are thankful to the Kluwer Academic Publishers for the initiative and assistance that allowed us ultimately to accomplish the goal with this project.

A.M. Fridman
M. Ya. Marov
R. H. Miller
June 2002

STOCHASTIC APPEARANCE IN STARS AND HIGH ENERGY SOURCES

G.S.BISNOVATYI-KOGAN

Space Research Institute, Russian Academy of Sciences, Moscow, Russia
(E-mail: gkogan@mx.iki.rssi.ru)

Abstract. Many phenomena such as stellar variability, stellar explosions, behaviour of different kind of X-ray and gamma-ray sources, processes in active galactic nuclei and other astrophysical objects show stochastic features. Brief descriptions of these phenomena are given in this review.

1. Introduction

The following phenomena in astrophysics are known to show stochastic features.

1. Star motion in stellar systems.
2. Active galactic nuclei (AGN).
3. Stellar oscillations.
4. X-ray and gamma-ray bursters; quasi-stationary X-ray sources, like Her X-1.
5. Combustion detonation front in type I supernovae, showing a fractal structure.
6. Magnetic dynamo in stars.
7. Motion of comets in Solar system.

Only the first item, star motion in stellar system, is investigated in details among papers presented in this meeting. In present paper we shall give a brief review of all the above items. In particular, for item 1, we consider a simple system, consisting of two self-gravitating intersecting shells with or without the central gravitating body. The shells consist of stars with the same parameters of elliptic orbits.

2. Two Shells around SBH

Spherically symmetric stellar cluster may be approximated by a collection of spherical self-gravitating shells consisting of stars having the same orbit parameters. Dynamical behaviour of a cluster in the shell approximation was first considered by Yangurazova and Bisnovatyi-Kogan (1984). Analysis of motion of two such shells consisting of stars with net radial motion with a reflecting inner boundary, done by Miller and Youngkins (1997), have shown a stochastic behaviour in their motion. More realistic model with two shells of stars moving along elliptical orbits, with and without the central gravitating body, was investigated by Barkov *et al.* (2001).

The motion of each star in two shells is characterized by the specific angular momentum J_1, J_2 which do not change during intersections, and energies, which are changing during intersections, and which initial values are

$$E_{1(0)} = \frac{m_1 v_{1(0)}^2}{2} - \frac{Gm_1(M + m_1/2 + m_2)}{r} + \frac{J_1^2 m_1}{2r^2}, \tag{1}$$

$$E_{2(0)} = \frac{m_2 v_{2(0)}^2}{2} - \frac{Gm_2(M + m_2/2)}{r} + \frac{J_2^2 m_2}{2r^2}. \tag{2}$$

Here $v = dr/dt$ is the radial velocity of the shell and $J^2 m/2r^2$ is the total kinetic energy of tangential motions of all stars of the shell, M is the mass of a central body. The term $m_1/2$ in (1) is due to the self-gravity of the shell. By the index (0) we mark the initial stage before the first intersection, when the shell '2' is inside the shell '1'. Note, that we use here the values of energies (negative), which determine the elliptical trajectories of stars in the shells, including the gravitational energy without normalization. Let shells intersect at a some radius $r = a_1$ and at some time $t = t_1$ after which the shell '1' becomes inner and shell '2' – outer. The 'energies' of the shells designated it by the index (1) than become:

$$E_{1(1)} = \frac{m_1 v_{1(1)}^2}{2} - \frac{Gm_1(M + m_1/2)}{r} + \frac{J_1^2 m_1}{2r^2}, \tag{3}$$

$$E_{2(1)} = \frac{m_2 v_{2(1)}^2}{2} - \frac{Gm_2(M + m_2/2 + m_1)}{r} + \frac{J_2^2 m_2}{2r^2}. \tag{4}$$

The matching conditions at the intersection point $r = a_1, t = t_1$ are written as:

$$E_{1(0)} + E_{2(0)} = E_{1(1)} + E_{2(1)};$$

$$v_{1(0)}(t_1) = v_{1(1)}(t_1); \qquad v_{2(0)}(t_1) = v_{2(1)}(t_1), \tag{5}$$

defining the conservation of the total energy of the system and continuity of the velocities through the intersection point. It follows than from (1)–(5):

$$E_{1(1)} = E_{1(0)} + \frac{Gm_1 m_2}{a_1}; \qquad E_{2(1)} = E_{2(0)} - \frac{Gm_1 m_2}{a_1}. \tag{6}$$

The relations (1)-(6) are valid for all subsequent intersections of the shells.

The shell motion in the Newtonian gravitational field is described by the algebraic relations. The motion of one shell is completely regular, but at presence of intersections the picture changes qualitatively. The shell intersections result in chaos in their motions. This chaos appears in the fully integrable system, described algebraically by integrals of motion. The origin of this chaos is different from

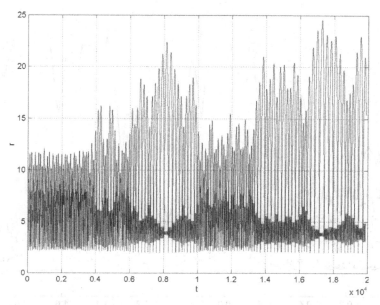

Figure 1. The chaotic motion $r_{1,2}(t)$ of shells with equal masses $m/M = 0.08$ on long time interval.

the chaotic behaviour of non-integrable orbits in non-axisymmetric gravitational potential (Merritt, 2001). Character of chaos in the shell motion depends, mainly, on the mass ratio of a shell and a central body. For small mass ratios the motion of the shells occurs basically in the field of the central body, and after the intersection there is a little change in a trajectory of each shell. However it is possible to observe randomness of their behaviour in Figure 1 from Barkov *et al.* (2001).

Calculations by Barkov *et al.* (2001) have shown, that very small variations in the initial parameters drastically change the picture of the oscillations, which is a characteristic for a chaotic behaviour. In the case of massive shells the exchange of energy between shells occurs more intensively. As a result we have the obviously expressed chaotic behaviour of shells presented in Figure 3 from Barkov *et al.* (2001) for the case $m/M = 0.15$.

The example of a chaotic behaviour of two intersecting self-gravitating shells, moving in their own gravitational field, without a central mass, is shown in Figure 4. Some other example of the chaotic motion of intersecting self-gravitating shells are given by Barkov *et al.* (2001).

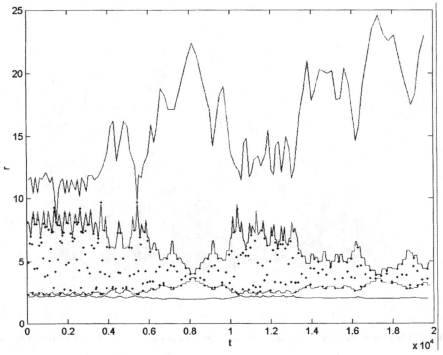

Figure 2. The result of treatment Figure 1. Here lines of the shell maxima (top), shell minima (bottom), maxima of the bottom shell trajectory (upper middle) and minima of the bottom shell trajectory (lower middle) are represented. Shell intersection places are marked by black points.

Ballistic mechanism of energy exchange between gravitating particles may be important in the formation of structures in a cold dark matter, where gravitational instabilities are developing in masses, much less than galactic ones, and gravitationally bound massive objects, consisting mainly from the dark matter may be formed.

3. Stars Around a Supermassive Black Hole in Active Galactic Nuclei

It is now widely believed that active galactic nuclei and quasars are radiating due to accretion of matter into a supermassive black hole, according to Lynden-Bell (1969) model. A supermassive black hole is surrounded by a dense stellar cluster. Its member stars may supply matter for the accretion, when they make a close approach to the central body. The fate of the star depends on the mass of the black hole. When $M_{bh} < 3 \times 10^7 M_\odot$, the solar-type stars are destroyed by tidal forces at radius $r_t = (M_{bh}/M_\odot)^{1/3} R_\odot$, and matter is forming an accretion disk. For black holes with mass greater than $3 \times 10^7 M_\odot$ tidal forces are not enough for disruption

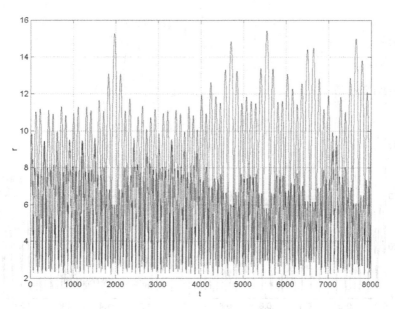

Figure 3. Chaotic shell oscillations with mass ratio $m/M = 0.15$.

of such stars, and they are falling directly into the black hole from the distance of $2 - 3$ the Schwarzschild gravitational radius $r_g = 2GM_{bh}/c^2$. This is because the radius of the tidal disruption r_t depends on the black hole mass as $r_t \sim M_{bh}^{1/3}$, and the radius of the gravitational capture $r_{grav} \sim M_{bh}$, so at large M_{bh} we have $r_t < r_{grav}$, and gravitational capture occurs before the tidal disruption happens (see e.g. Bisnovatyi-Kogan *et al.*, 1980). The tidal disruption of stars may be an important supply of matter in low-luminosity AGN (Rees, 1994).

The region in the phase space from which stars may be disrupted or absorbed by the black hole occupies a small region called loss cone. In the non-collisional stellar cluster this cone is almost empty, and it is filled only due to rare collisions which provide a diffusion of stars into a loss cone. This process may be very slow, and sometimes cannot give enough matter to explain the most luminous objects. The situation may be more optimistic when stars around the black hole move along chaotic orbits due to the particular shape of a gravitational potential. Filling of the loss cone, according to Norman and Silk (1983), occurs not only by slow diffusion, but also by stars entering the loss cone during the motion along chaotic orbits. The latter may even be more effective.

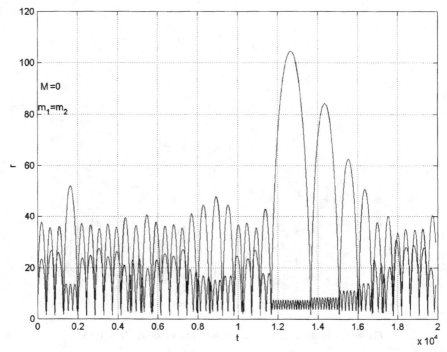

Figure 4. The chaotic motion $r_{1,2}(t)$ of two self-gravitating shells in their own gravitational field without a central mass.

4. Stellar Oscillations

Stars show all kinds of variability: from periodic to totally irregular. The variability arises from the influence of the thermal processes and displays in their dynamic behaviour. Irregular variability is observed in stars with large convective envelopes. The most intensive convection is developed in the envelopes of cool stars, where matter opacity is high, and radiative gradient exceeds the adiabatic one. This is related to stars with masses 1–2 $M_{\odot}$ in the stage of gravitational contraction (T Tauri stars), and to low-mass stars with $M \leq \sim 0.3 M_{\odot}$, which remain almost fully convective during all their life. Both type of stars show irregular variability in the form of long and short flares. Short time variability of of several T Tauri type star is represented in Figure 5 from the paper of Kuan (1976); the light curve of T Tauri type star DF Tauri in different time scales is shown in Figure 6 from the paper of Zaitseva and Lyutyi (1976).

Detailed description of chaotic flares in low mass UV Ceti stars is given in the book of Gershberg (1978). Light curves in Figures 5 and 6, show an evident chaotic behaviour.

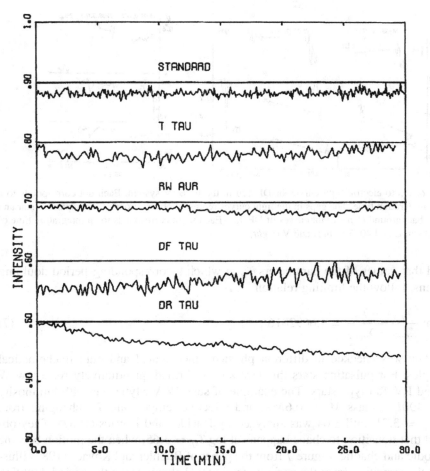

Figure 5. Time scans of the ultraviolet intensities of a standard star and several T Tauri stars obtained on 1976 January 26 (UT). The integration time for the standard is 5 s and for the rest 10 s. The intensities have been corrected for atmospheric extinction and are in arbitrary unit.

Another type of chaos in stellar light curve appears in pulsating stars in which the regular pulsations are transforming into the chaotic ones in the course of stellar evolution. The transition from regular pulsations to the chaotic regime occurs with change of a characteristic during the evolution, in agreement with the mechanism of period doubling (Feigenbaum, 1983). The behaviour becomes purely chaotic when the parameter λ is approaching the limiting value λ_{lim}. The universal law describing the approach to chaos was discovered by Feigenbaum (1983). He had

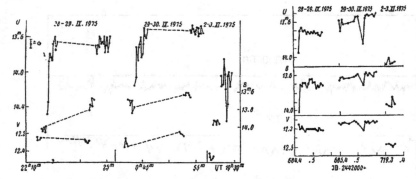

Figure 6. Photo-electric light curves of DF Tau in the U, B, V system. Each dot corresponds to a 10-sec integration time. The gaps in the light curves represent observations of a comparison star and the sky background (*left*). Light curves of DF Tau. Each point corresponds to an integration time of 80–100 s in U and 40–50 s in B and V (*right*).

found that values of the parameters λ_n at which a corresponding period doubling happens, follow the limiting relation

$$\lim \frac{\lambda_{n+1} - \lambda_n}{\lambda_{n+2} - \lambda_{n+1}} = 4.6692016.... \tag{7}$$

This law was checked in different physical, mechanical and pure mathematical examples. For pulsating stars this law was confirmed quantitatively for a few W Vir and RV Tau type stars. The example of such W Vir type star with luminosity $L = 400L_\odot$, mass $M = 0.6M_\odot$, and effective temperature T_{ef} changing from $\log T_{ef} = 3.71$ until 3.64 was analyzed by Buchler and Kovacs (1987). They obtained that transition to chaos happens at $\log T_{ef} = 3.65$, when the motion became aperiodic and chaotic. Figure 7 from the paper of Buchler and Kovacs (1987) illustrates the transition from the periodic to chaotic behaviour via the period doubling mechanism.

5. High-energy Sources

Irregular variabilities are observed in most X-ray and gamma-ray sources. Strong X-ray sources consist of a neutron star or a black hole in a binary system, and the main energy supply in these objects comes from accretion into the compact star from the companion. In the case of a black hole the main source of radiation is an accretion disk, which is formed due to high angular momentum of the falling matter. In the case of a neutron star, in addition to the accretion disks, important processes occur on the surface of the neutron star, and in its magnetosphere, when the star is strongly magnetized. Both the accretion disk and the neutron star surface, as well as the magnetosphere suffer from different kind of instabilities,

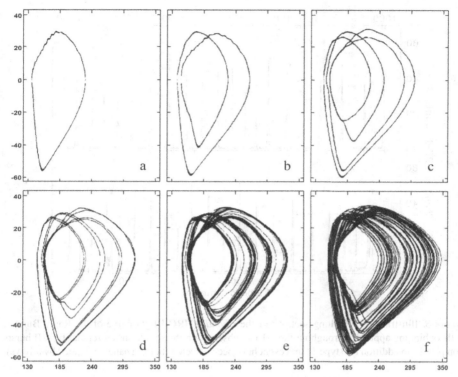

Figure 7. Plots of v_* (ordinate) - R_* (abscissa) for models a trough f (the asterisk denotes the 55th zone of the models); R_* in 10^{10} cm, v_* in km s^{-1}. The logarithms of effective temperature T_{ef}, in K, $\log T_{ef}$ (and corresponding periods) are the following: a – 3.71 (11 587); b – 3.69 (26 196); c – 3.67 (59 040); d – 3.66 (125.784); e – 3.65 (∞); f – 3.64 (∞).

leading to irregular radiation flux. The instabilities in the accretion disk of visco-thermal origin determine short-time fluctuations on scales from seconds to milli-seconds, observed in most X-ray sources, containing black holes (Cherepashchuk, 1996). Accretion disk instabilities, determining transition between quasi-laminar and highly turbulent states are probably responsible for appearance of soft X-ray transient (X-ray novae) (see review of Cherepashchuk (2000)). Similar instability may also explain cataclismic variables (see Spruit and Taam, 2001), where flashes occur with intervals of few months. These sources are also binary systems, containing white dwarf and low-mass star supplying matter for an accretion disk formation around the white dwarf. The energy of these flashes is several orders of magnitude less than in X-ray novae.

Interaction of matter with stellar surface and stellar magnetosphere determine chaotic features in radiation of the sources, containing neutron stars. Voges *et al.* (1987) have found irregular variability in the form of 'deterministic chaos' in the

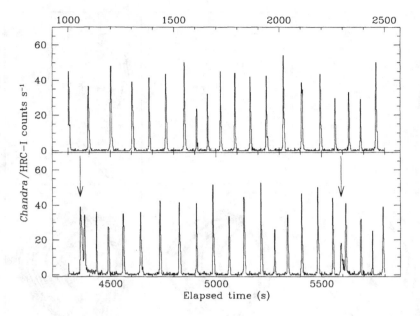

Figure 8. Illustrative 1500s-long sections of the *Chandra*/HRC-I light curve of the Rapid Burster, with 1s binning applied. Throughout the ~13 ks observation the source shows regular type-II bursts (top panel); in addition two type-I burst event have been detected (bottom panel, indicated by arrows).

variations of the pulse shape of the famous X-ray pulsar Her X-1. It is probably connected with instabilities in the accretion flow inside the magnetosphere.

Matter falling into the neutron star becomes degenerate soon after joining its envelope, when pressure almost does not depend on the temperature. In this conditions the thermal instability of thermonuclear burning develops, leading to appearance of X-ray bursters, observed in low-mass X-ray binary systems. This bursts occur non-periodically and show chaotic properties. Example of chaotic flares in the unique famous burster source, called 'rapid burster' is represented in Figure 8 from the paper of Homer *et al.* (2001) using observations of Chandra X-ray satellite.

6. Fractal Structure of Detonation Front in SN I

Supernovae explosions are the most spectacular astronomical events, by which massive stars end their life. Type I SN appears due to thermonuclear explosion of a degenerate $C - O$ stellar core with a mass equal to the Chandrasekhar mass limit $\sim 1.4\,M_\odot$. The thermal instability starts in the central part of the core, and the flame propagates outside to the surface of the core. Physical and astronomical analysis of the problem led to conclusion, that the flame propagation should be highly unstable, and the surface of the front of the flame should be highly non-smooth, with a fractal properties. Development of a fractal structure of the flame front strongly enhances the surface of the burning, increasing the rate of the energy production. That may change the regime of the flame propagation from deflagration, when the flame front moves with a sub-sonic speed, to detonation, in which this front moves with the speed equal or larger than the local sound speed. The necessity of this transition follows also from the interpretation of observational data on the isotope distribution of different elements, produced by ejection of matter during SN I explosions. Theoretical analysis have shown existence of two types of instabilities in the flame front: Landau-Darrieus (Blinnikov and Sasorov, 1996), and Rayleigh-Taylor (RT) instability (Niemeyer and Hillebrandt, 1995). 2-D numerical calculations made by Niemeyer *et al.* (1996) have shown a developement of RT instability, which appeares in many scales, approaching the fractal structure (see Figure 9).

The simulations involved an Eulerian PPM-based code to solve the 2-D hydro-dynamical equations. All calculations were based on a 2-D stationary grid with 256×64 zones in spherical (r, ϑ) coordinates. Assuming rotational and equatorial symmetry, the boundary conditions were chosen to be reflecting everywhere except at the outer radial edge, where outflow was allowed. The initial model represented a white dwarf with a mass equal to the mass of the Chandrasrkhar limit, a central density $\rho_c = 2.8 \times 10^9$ g cm^{-3} and a central temperature $T_c = 7 \times 10^8$ K. At $t \approx 0.8$ s, displayed in Figure 9, the RT-instability of the burning front has developed to its maximum extent. The turbulent flame speed is now $u(\Delta) \approx 2 \times 10^7$ cm s^{-1} at the points of maximum turbulent sub-grid energy. Four major bubbles can be identified, separated by thin streams of unburned C+O-material.

7. Stellar Dynamos

Dynamo models are used for explanation of the origin of stellar and planet magnetic fields. The irregular changes in the Earth magnetic field polarity with intervals $\Delta = (0.1 - 20) \times 10^6$ years have been explained by the chaotic behaviour of the dynamo process. The famous example of the chaotic dynamo is the toy two-disk Rikitake model (Cook and Roberts, 1970). This model consists of two magnetically connected disks (see Figure 10) with the following given parameters:

G – momentum of force, Ω – angular velocity, I – electrical currant, R – resistivity, L – self-inductance, M – mutual inductance, C – momentum of inertia.

The equation describing the behaviour of this system (mechanical and magnetic) are written as

$$L_1 \frac{dI_1}{dt} = -RI_1 + M\Omega_1 I_2, \qquad C_1 \frac{d\Omega_1}{dt} = G_1 - MI_1 I_2 \qquad (8)$$

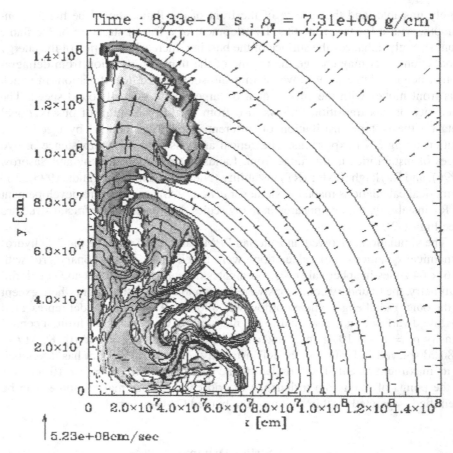

Figure 9. Core region in the SNI-model at maximally developed RT instability Density contours are separated by $\delta\rho = 3 \cdot 10^7$ g cm^{-3}, the maximum energy generation rate (shaded regions) is $\dot{S}_{max} = 1.3 \cdot 10^{19}$ ergs g^{-1} s^{-1}.

for the first disk, and corresponding equations for the second disk. Taking identical disks under the same external action G, we find two values of time, characterizing the system:

$$\tau_m = \frac{CR}{GM}, \quad \text{and} \quad \tau_l = \frac{L}{R}. \tag{9}$$

Here τ_m estimates the acceleration time, and τ_l characterizes the magnetic field damping. The ratio of these times $\mu^2 = \tau_M/\tau_l$ is the non-dimensional parameter dividing stochastic behaviour ($\mu \geq 1$) from the regular oscilaltions.

There are indications that the solar 22 year cycle is also showing stochastic behaviour on long-term period, and could be considered as a coherent strange attractor (Ruzmaikin *et al.*, 1992).

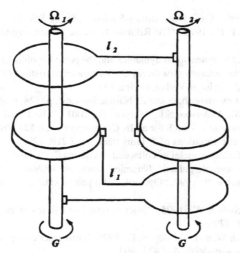

Figure 10. A schematic picture of Rikitake dynamo.

8. Chaos in Solar System

Appearance of most comets in the region of visibility is unpredictable. Theoretical analysis of Sagdeev and Zaslavsky (1987) had shown, that gravitational influence of Jupiter and Saturn creates a layer around the Sun in which orbits of comets become stochastic. Diffusion of the orbit parameters of comets in the stochastic layer make them to enter randomly the region when they become visible.

Acknowledgements

The work was partially supported by RFFI grant 99-02-18180, INTAS-ESO grant No.120 and INTAS grant 00-421.

References

Barkov, M. V., Belinski, V. A., Bisnovatyi-Kogan, G. S.: 2001, 'Chaotic motion and ballistic ejection of gravitating shells', astro-ph/0107051.

Bisnovatyi-Kogan, G. S., Churayev, R. S. and Kolosov, B. I.: 1980, 'A star cluster around a massive black hole: a steady-state numerical solution', *Sov. Astron. Letters* **6**, 82.

Blinnikov, S. I., Sasorov, P. V.: 1996, 'Landau-Darrieus instability and the fractal dimension of flame fronts', *Physical Review E* **53**, 4827.

Buchler, J. R., Kovacs, G.: 1987, 'Period doubling bifurcations and chaos in W Virginis model', *ApJ Letters* **320**, L57.

Cherepashchuk, A. M.: 1997, 'The Masses of Black Holes in X-ray Binary Systems', *Astrophysics and Space Science* **252**, 375.

Cherepashchuk, A.M.: 2000, 'X-ray Nova Binary Systems', *Space Science Reviews* **93**, 473.

Cook, A. E. and Roberts, P. H.: 1970, 'The Rikitake two-disc dynamo System', *Proc. Camb. Phyl. Soc.* **68**, 547.

Feigenbaum, M.: 1983, 'Low dimensional dynamics and the period doubling scenario', in *Dynamical systems and chaos; Proceedings of the Sitges Conference on Statistical Mechanics, Sitges, Spain, September 5–11, 1982*, Springer-Verlag, Berlin, 131.

Gershberg, R. E.: 1978, 'Low-mass flare stars', Nauka. Moscow, p. 128, in Russian.

Homer, L., Deutsch, E. W., Anderson, S.F., Margon, B.: 2001, 'The Rapid Burster in Liller 1: the Chandra X-ray Position and a Search for an IR Counterpart', *AJ*, **122**, 2627.

Kuan, P.: 1976, 'Photometric variations of T Tauri stars', *ApJ* **210**, 129.

Lynden-Bell, D.: 1969, 'Galactic nuclei as collapsed old quasars', *Nature* **223**, 690.

Merritt, D.: 2001, 'Non-integrable galactic dynamics', astro-ph/0106082.

Miller B. N. and Youngkins V. P.: 1997, 'Dynamics of a pair of spherical gravitating shells', *Chaos* **7**, 187.

Niemeyer J. C. and Hillebrandt W.: 1995, 'Microscopic instabilities of nuclear flames in Type Ia supernovae', *ApJ* **452**, 779.

Niemeyer J. C., Hillebrandt W. and Woosley, S. E.: 1996, 'Off-center deflagrations in Chandrasekhar mass Type Ia supernova models', *ApJ* **471**, 903.

Norman, C. and Silk, J.: 1983, 'The dynamics and fueling of active nuclei', *ApJ* **266**, 502.

Rees, M. J.: 1994, 'Models for variability in AGNS', in *Proc. IAU Symposium no. 159*, 239.

Ruzmaikin, A., Feynman, J. and Kosacheva, V.: 1992, 'On Long-Term Dynamics of the Solar Cycle', in *The solar cycle; Proceedings of the National Solar Observatory/Sacramento Peak 12th Summer Workshop, ASP Conference Series*, ASP: San Francisco **27**, 547.

Sagdeev, R. Z., Zaslavsky, G. M.: 1987, 'Stochasticity in the Kepler Problem and a Model of Possible Dynamics of Comets in the Oort Cloud', *Nuovo Cimento B* **97**, 119.

Spruit, H. C. and Taam, R. E.: 2001, 'Circumbinary Disks and Cataclysmic Variable Evolution', *ApJ* **548**, 900.

Voges, W.; Atmanspacher, H.; Scheingraber, H.: 1987, 'Deterministic chaos in accreting systems – Analysis of the X-ray variability of Hercules X-1', *ApJ* **320**, 794.

Yangurazova, L. R., Bisnovatyi-Kogan, G. S.: 1984, 'Collapse of spherical stellar system', *Ap. Sp. Sci.* **100**, 319.

Zaitseva, G. V. and Lyutyi, V. M.: 1976, 'Photometry with 10-sec resolution for the T Tauri variable DF Tauri', *Sov. Astro. Lett.* **2**, 167.

OBSERVATIONAL MANIFESTATIONS OF PRECESSION OF ACCRETION DISK IN THE SS 433 BINARY SYSTEM

ANATOL CHEREPASHCHUK

Sternberg Astronomical Institute, 13, Universitetsky Prospect, Moscow, 119992, Russia

Abstract. Basic properties of the unique object SS 433 are described. Observational spectroscopic and photometric manifestations of a precessing accretion disk around a relativistic object in this X-ray binary system are presented.

1. Introduction

Object SS 433 is the unique eclipsing X-ray binary system containing supercritical precessing/slaving accretion disk and collimated relativistic jets with velocity of streaming gas $\sim 0.26\,c$ (see e.g. Catalogue by Cherepashchuk *et al.*, 1996, and references therein). This object has been investigated by many authors since 1978 (e.g. Clark and Murdin, 1978; Margon *et al.*, 1979; Mammano *et al.*, 1980; Milgrom, 1979; Fabian and Rees, 1979; Crampton *et al.*, 1980, van den Heuvel *et al.*, 1980; Cherepashchuk, 1981a; Crampton and Hutchings, 1981; Margon, 1984; Cherepashchuk, 1988; Gladyshev *et al.*, 1987; Kemp *et al.*, 1986; Stewart *et al.*, 1987; Vermeulen *et al.*, 1993; Kotani *et al.*, 1996; Kotani, 1998; Goranskii *et al.*, 1998a,b; Panferov and Fabrika, 1997; Marshall *et al.*, 2001).

The main characteristic feature of SS 433 is the presence of highly collimated relativistic ejecta–jets, which precess with a period of $162.^d5$ and which are related to the central parts of an apparently supercritical accretion disk around the relativistic object. The precession period of $162.^d5$ is observed both in the Doppler shifts of moving H, HeI, FeXXV and Fe XXVI emission lines and in the optical light variations of the system. The shape of the orbital optical light curve changes greatly with precession phase. The orbital period of $13.^d082$ exibits no variations despite the intense mass outflow through stellar wind ($v \approx 3000$ km/s, $\dot{M} \simeq 10^{-4}\,M_\odot$/year) both from the optical star and from the supercritical accretion disk. The light variations of SS 433 are also characterized by optical flares on time scales of several days, which are correlated with radio flares and pass ahead of them by one or two days (Gladyshev *et al.*, 1983, Cherepashchuk and Yarikov, 1991).

The main puzzles of SS 433 remain to be solved: the nature of the relativistic object (a neutron star or a black hole), the mechanisms of collimation and acceleration of jet matter to relativistic velocities of 80000 km/s, nature of the precession phenomena in the binary system. In our review we describe basically observational

Space Science Reviews **102**: 23–35, 2002.
© 2002 *Kluwer Academic Publishers. Printed in the Netherlands.*

properties of the precessional variabilities in SS433 in connection with possible observational appearances of deterministic chaos in this unique binary system.

2. Basic Characteristics of SS 433 Binary System

SS433 is an eclipsing X-ray binary system consisting of a normal optical star filling (or overfilling) its Roche Lobe and a relativistic object surrounded by the supercritical geometrically thick precessing accretion disk. Three systems of emission lines are observed in the spectrum of SS433. The positions of stationary lines are close to their laboratory positions, two other systems of emission lines are moving along the spectrum with the period of 162.5 days and the amplitude $\sim$ 900 Å. Moving emission lines and the period about $162.^d5$ for them were discovered by Margon *et al.* (1979) (see the review of Margon, 1984 and references therein). According to Milgrom (1979) and Fabian and Rees (1979) moving emission lines are formed in collimated jets with velocity of streaming matter $\sim$ 80000 km/s. The both jets are precessing with the period $162.^d5$. The angle between the direction of jets and the precession axis is $\sim$ 20° but the angle between the line of sight and the precession axis is close to 79°. Due to Doppler effect the moving emission lines are regularly shifted along the spectrum of SS 433. Values of red and blue shifts of moving emission lines $z = \frac{\lambda - \lambda_0}{\lambda}$ are described by the well known relativistic relation:

$$ z = \left(1 - \frac{v^2}{c^2}\right)^{-1/2} \left(1 \pm \frac{v}{c}\cos\psi\right) - 1, \tag{1} $$

where $v = 80000$ km/s is the velocity of streaming matter in the jets, c – the velocity of light, ψ – the angle between the line of sight and the direction of jets.

Discovery of optical eclipses in SS 433 (Cherepashchuk, 1981) as well as discovery of periodic ($p \simeq 13.^d1$) radial velocity variability from the measurements of stationary H_α, H_β and HeII 4686 lines (Crampton *et al.*, 1980; Crampton and Hutchings, 1981) allowed us to conclude that relativistic jets in SS 433 binary system are perpendicular to the plane of supercritical accretion disk which precesses with the period $\sim 162.^d5$. Therefore, the origin of collimated relativistic jets in SS 433 is closely related to the processes of disk accretion of matter onto the relativistic object. Precession of jets is due to precession of the accretion disk in SS 433 binary system.

From the analysis of optical and X-ray observations the following characteristics of SS 433 binary system have been determined.

The SS 433 eclipsing binary consists of an optical star filling or overfilling its Roche lobe and a relativistic object (a neutron star or a black hole) surrounded by the precessing geometrically thick supercritical accretion disk. Due to overfilling Roche lobe by the optical star there is intense secondary mass exchange in the system. The lines of the optical star are not observed in the summary spectrum of SS 433, therefore the precise spectral class of the optical star is unknown. The mass function of the optical star for SS 433 is also unknown. From the observations of emission line HeII 4686 which presumably is formed on the central parts of the accretion disk the mass function of the relativistic object $f_x(m)$ has been measured. The results concerning mass function of the relativistic object obtained by different authors contradict each other:

$$f_x(m) = \frac{m_v^3 \sin^3 i}{(m_x + m_v)^2} = \begin{cases} 5 \div 10 M_\odot & \text{(Crampton and Hutchings, 1981)} \\ 7.7 M_\odot & \text{(Fabrika and Bychkova, 1990)} \\ 2 M_\odot & \text{(D'Odorico } \textit{et al.}, 1991). \end{cases}$$

Here m_v is the mass of the optical star, m_x – that of the relativistic object, i – inclination of the orbital plane of the binary system. The analysis of optical eclipses in SS 433 (Antokhina and Cherepashchuk, 1987) leads to the estimate of the mass ratio of the components $q = m_x/m_v > 0.25$. On the other hand, the analysis of X-ray eclipses in SS433 (Antokhina et al., 1992; Kotani, 1998) enables us to obtain the estimate $q < 0.25$. Thus, estimates of q obtained from optical and X-ray eclipses are overlapped at $q = 0.25$. Using this value of q and the relativistic object mass function $f_x(m) = 10.1 M_\odot$ (Crampton and Hutchings, 1981) we obtain the relativistic object mass $m_x = 4 M_\odot$ (a corresponding value of $m_v = 16 M_\odot$) which favours the presence of a black hole in SS 433 binary system (Antokhina and Cherepashchuk, 1987). Note that the large value of $f_x(m) = 7.7 M_\odot$ is confirmed by observations obtained with 6-meter telescope of Special Astrophysical Observatory (Fabrika and Bychkova, 1990). At the same time a small value $f_x(m) = 2 M_\odot$ is obtained by D'Odorico et al. (1991) on the basis of new high resolution spectroscopic observations of the HeII 4686 emission line in SS433. At $q = 0, 25$ this corresponds to masses $m_x = 0.8 M_\odot$ and $m_v = 3.2 M_\odot$. With such characteristics the system SS433 becomes analogues to the X-ray binary system Her X-1/HZ Her containing a moderate–mass optical star and a neutron star. In this case it is difficult to explain a huge luminosity of the optically bright accretion disk ($\sim 10^{39} \div 10^{40}$ erg/s) and a high mass loss from the optical star ($\sim 10^{-4} M_\odot$/year). It should be noted, however, that the spectroscopic observations by D'Odorico et al. (1991) were taken at the precession period phases far from the moment of the maximum separation of moving emission lines in the SS 433 spectrum. As was shown by Fabrika and Bychkova (1980), it is at the moment of moving emission maximum separation (the most part of the accretion disk is exposed to a terrestrial observer) when the mass function $f_x(m)$ for SS 433 is maximal. Thus, the problem of the black hole existance in the system SS 433 still waits for its ultimate solution until new high-resolution spectroscopic observations taken at phases of the maximum separation of moving emissions will become available. Let us note that small values of $q = 0.15 \div 0.25$ are inconsistent with the optical observations of SS 433. Since the inclination of the orbital plane for SS 433 is fixed ($i \simeq 79°$) and the optical star fills its Roche Lobe, for small values of $q = 0.15 \div 0.25$ total eclipses of the precessing accretion disk by the optical star should be observed at all phases of the precessional period ($p = 162.^d5$) which is not the case.

From optical and X-ray observations the following characteristics of collimated relativistic jets and the accretion disk were determined (Borisov and Fabrika, 1987; Kotani, 1998):

- the length of the jet in the optical range $l_{opt} = 2 \cdot 10^{14} - 3 \cdot 10^{15}$ cm,
- the opening angle of the jet $\Delta\phi = 1.°0 \div 1.°4$,
- the length of the jet in X-ray $l_x = 2 \cdot 10^{13}$ cm,
- the temperature of the streaming plasma at the base of the jet $T_{base} = 20$ keV,
- the density $n_e^{base} = 2.5 \cdot 10^{12}$ cm^{-3},
- the mass loss rate through the jets $\dot{M} = 4 \cdot 10^{-6} M_\odot$/year.

Recently Marshall et al. (2002) presented improved values of these parameters of jets based on new X-ray observations of SS 433 from Chandra satellite.

A value of $\dot{M}$ for quasispherical non-relativistic wind from accretion disk ($v = 3000$ km/s) is close to $\dot{M} = 10^{-4} M_\odot$/year. If the radius of a relative orbit of the system $a = 4 \cdot 10^{12}$ cm, the value of kinetic power of matter emanated through the jets is $L_k = 10^{40}$ erg/s. This value of L_k is close to the value of bolometric luminosity of optically bright precessing accretion disk, $L_{opt} = 10^{39} \div 10^{40}$ erg/s ($T_{disk} \approx 30000 - 40000$ K). X-ray luminosity of SS 433 is $\sim 10^{36}$ erg/s and is only $\sim 10^{-3} \div 10^{-4}$ of bolometric luminosity of the accretion disk. There is a strong evidence that the optically bright accretion disk in SS 433 is in supercritical regime of accretion. First description of a supercritical accretion disk has been published by Shakura and Sunyaev (1973). Marshall et al. (2002) obtained a new value of the kinetic power for jets, $L_k = 3.2 \cdot 10^{38}$ erg/s.

3. Possible Mechanisms of Accretion Disk Precession in SS 433

There are two basic mechanisms proposed for accretion disk precession in SS 433 binary system: those of driven precession (Katz, 1973, 1980, 1975; Petterson, 1975) and of slaved precession (Roberts, 1974; van den Heuvel et al., 1980; Cherepashchuk, 1981b,c; Bisikalo et al., 1999).

According to Shakura (1972), if the plane of accretion disk is tilted relatively to the orbital plane of a binary system the disk can precess.

Driven precession of an accretion disk was investigated by Katz for the X-ray binary system HZ Her=Her X1 (Katz, 1973) and for SS 433 object (Katz, 1980). According to Katz (1997), an accretion disk inclined to the orbital plane of a binary system at the angle of θ_0 will precess at the rate of

$$\Omega_0 = -\frac{3}{4} \frac{Gm_2}{a} \left(\frac{a_d}{a}\right)^2 \frac{\cos\theta_0}{(Gm_1a_d)^{1/2}}, \tag{2}$$

where m_1 is the mass of an accreting object, m_2 – the mass of its companion, a – their separation, a_d – the disk radius. Ω_0 is a measure of the torques that drive the nodding motions even if the actual precession has other contributions and occurs at a different rate. This Newtonian-driven precession is usually much faster than geodetic precession. In the case of SS 433 the precession frequency Ω_0 is distinguished from the observed precession frequency ω_{prec}: $\Omega_0/\omega_{prec} \approx 2.1$.

Inclination of the accretion disk to the orbital plane may be due to self-induced radiation-driven warping of the accretion disk (Pringle, 1996; Maloney et al., 1996; Maloney and Begelman, 1997). Centrally illuminated accretion disks are unstable to warping because of the pressure of reradiated flux which, for a nonplanar disk, is nonaxisymmetric and therefore exerts a torque. Such a warped accretion disk may precess due to driven precession.

Katz et al. (1982) showed that the precessing disk in SS 433 binary system is expected to perform a kind of nodding motion with the frequency of $2f_{13} + 2f_{162}$ driven by the gravitational torque of the optical star. They also showed that the inclination angle of the nodding disk, relative to a fixed line of sight, is modulated mainly with the frequency $2f_{13} + f_{162}$. Their calculations predict that the amplitude at this frequency could be greater by a factor of 10 than the amplitude at any other beat frequency (see also Newson and Collins, 1982).

The model of a slaved accretion disk was proposed by Roberts (1974). In this model, short residence time for matter in the disk permits it to follow the precessional motion of a misaligned companion star. Katz et al. (1982) claimed that the observed amplitude of the nodding motions strongly favors slaved precession over driven one as the dominant process in SS 433 binary system. Misalignment of the vector of the angular momentum of a companion star and that of the total angular momentum of a binary system may be due to asymmetric supernova explosions in close binary systems (Roberts, 1974; Cherepashchuk, 1981b,c). Precession of angular momentum of the optical star is due to tidal torques from the relativistic object. Matese and Whitmire (1982) considered the nutation of the optical star and its impact on a slaved disk. Collins (1985) studied the general case of nutation of a spinning object in a binary system.

According to Margon (1984), early suggestions (Martin and Rees, 1979, Sarazin et al., 1980) concerning relativistic Lense–Thirring effect as the main cause of precession in SS 433 can be ruled out because such a scheme requires a highly compact accretion disk contrary to the inferences from the photometric data and nodding motions of the accretion disk which need highly extended disk. Recent 3–dimensional gas–dynamic calculations (Bisicalo et al., 1999) confirmed the model of a slaved disk in interacting binary systems.

4. Spectroscopic Manifestations of the Precessing Accretion Disk

First of all, let us summarize characteristics of precessional variability obtained from spectroscopic observations of moving emission lines in SS 433 binary system. Basic spectral variability is 162.5-day precessional motions of emission lines in the spectrum of SS 433 which are well described in the framework of a so called simple kinematic model (Margon, 1984). Baykal *et al.* (1993) estimated the epochs of the phases of maximal red and blue beam Doppler shifts from various cycles of the period in SS 433 with the use of an updated optical spectroscopic database which covers a time span of over 13 years. By constructing power density spectra of phase residuals based on these epochs (residuals measured relative to a hypothetical clock with a constant period) these authors were able to characterize the statistical behavior of the clock noise in both red and blue beam $162.^d5$ cycles as a white noise process in the first time derivative of $162.^d5$ phase fluctuations. These features of the $162.^d5$ clock noise in SS 433 show a strong quantitative similarity to the clock noise of the 35^d periodicity of X-ray turns-on in HerX-1, lending further credence to the notion that the relevant underlying physical process in both binary systems is disk precession. An important quantity for a clock process chracterizable as white noise in frequency is the clock stability. This stability is parametrized by a dimensionless quality factor

$$Q = \frac{f}{\left(\Delta f^2\right)^{1/2}}, \tag{3}$$

where the denominator is the rms frequency increment per cycle. For SS 433 the value of $Q \approx 75$, for Her X-1 $Q \approx 37$ (Baykal *et al.*, 1993). These values of Q are in contrasts with orbital periods or geodetic precession periods which should either be good clocks with very high Q or show monotonically decreasing periods if dissipative processes shrink the orbit (Katz, 1997).

Although a simple kinematic model for SS 433 is remarkably successful, there are at least three separate types of significant deviations from this simple model (Margon, 1984). First type of deviations is presence, in $162.^d5$ periodicity of Doppler shifts of moving emission lines, of a short-term (for example, $6.^d28$) regular small amplitude ($\sim 5 \div 10\%$ of the major $162.^d5$ variations) deviations. These short term regular periodicities, which reflect effects of nodding motions (driven nutation) of the accretion disk, are due to tidal torques from the optical star.

The second type of the three deviations from the simple kinematic model are long-term (~ 1–2 months) irregular deviations of radial velocities for moving emission lines from average radial velocity curves (up to several thousands of km/s). As was pointed out above these deviations are distributed as white noise with dimensionless quality factor $Q \approx 75$.

The third type of deviations from the simple kinematic model is so called 'jitter' deviations. These appear to be random departures from predicted ephemeris of up to several thousand km/s on short timescales of one to a few days. This jitter effect is unlikely to be related to the accretion disk nodding motion and can be explained as a result of stochastic deviations in the beam pointing direction on a timescale of days with a characteristic angle of about 1–2 degrees which is close to the beam opening angle.

5. Photometric Manifestations of the Precessing Accretion Disk

Now we are going to present the results of optical observations of SS 433 which have been carried out in Sternberg Astronomical Institute (SAI) for the latest 20 years.

Photoelectric WBVR and spectroscopic CCD observations were obtained at 1.25 m and 0.5 m telescopes at the Crimean station of SAI and at 1 m telescope of Tyan-Shyan observatory. Photographic BV observations of SS 433 were carried out in Moscow and at the Crimean station of SAI. The total number of V observations of SS 433 including our data and all published data reaches $\sim$ 2500 in the V band. During the years 1995–2000 about 150 spectroscopic CCD observations of SS 433 in the H_α line region were obtained too, as well as values of Doppler shifts of moving emission lines H_α^+ and H_α^-.

The results of our observations and their analysis are presented in the papers by Goranskii, Esipov and Cherepashchuk (1998a,b). Here we describe the results published in these papers as well as some new results obtained in the last years.

In Figure 1–3 results of convolution of photometric observations of SS 433 with the precessional, orbital and nutational periods are presented. All the three periodicities are certain to be present in the photometric variability of SS 433. Optical bursts are also seen in the convolved light curves.

In Figure 4 results of frequency analysis for all photometric data for SS 433 are presented. Three peaks in Figure 4 correspond to the precessional, orbital, and nutational periods:

- $P_{prec} = 162.^d5(2)$
- $P_{orb} = 13.^d0821(2)$
- $P_{nut} = 6.^d2877(3) \approx (2f_{orb} + f_{prec})^{-1}$.

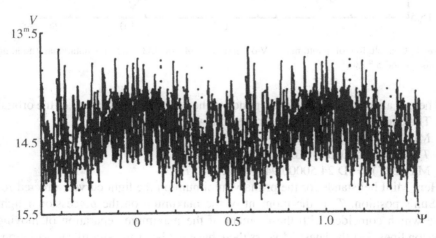

Figure 1. Convolution of photometric V-observations of SS 433 with the precessional period $162.^d5$.

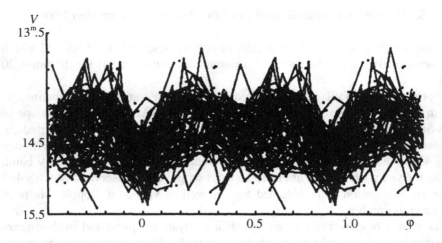

Figure 2. Convolution of photometric V-observations of SS 433 with the orbital period 13.^{d}082.

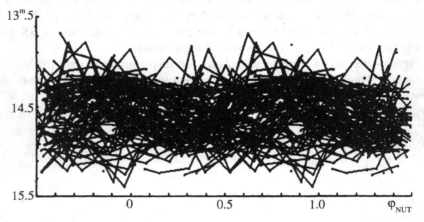

Figure 3. Convolution of photometric V-observations of SS 433 with the nutational (nodding) motions) period 6.^{d}286.

There is also the peak that corresponds to the period which is a half of the orbital one. The corresponding elements are as follows:

– MinI hel = JD 24 50023.62+13.^{d}08211(8) $\cdot$ E,

– T_3 = JD 24 50000 + 162.^{d}15 $\cdot$ E,

– Max(ΔV) = JD 24 50000.94 + 6.^{d}2877 $\cdot$ E.

Here MinI hel stands for the primary minimum on the light curve corrected for the Sun's position, T_3 is the moment of the maximum on the precessional light curve which coincides with the moment of the maximum separation of moving emission lines. For the latest 34 years there have not been any significant variations

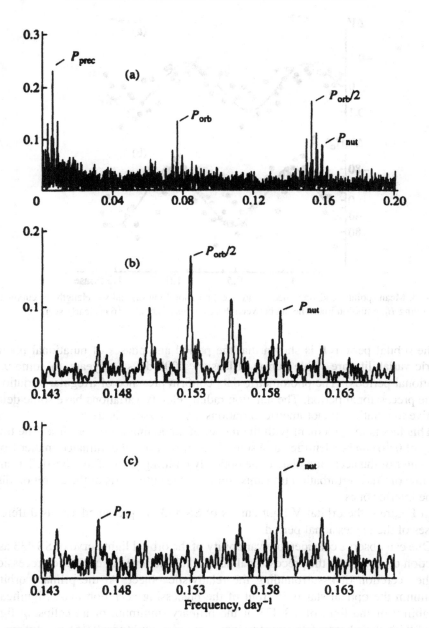

Figure 4. Power spectra of the photometric series of observations of SS 433 (a) the power spectrum in the range 5–5000^d for the original series of observations; (b) a portion of the power spectrum in the range 6–7^d after subtracting the precession period, which exhibits combs of peaks of the nutation period and the first harmonic of the orbital period; (c) the same portion of the power spectrum (as in Figure 4b) after subtracting the mean curve of the orbital period, which, in addition to a comb of peaks of the nutation period, exhibits a smaller comb of peaks of the second interaction period (P_{17}). The semiamplitude of the harmonic component is along the Y axis.

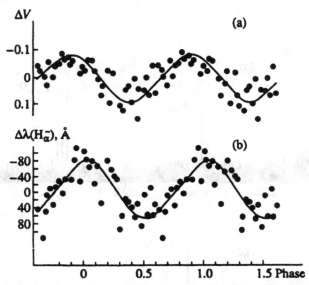

Figure 5. Mean (points) and smoothed (solid lines) nutation light (a) and wavelength (b) curves for the moving H_α emission line. The lag between these curves of about $0.^d6$ is clearly seen.

of the orbital period. It is shown that the period and phases of nutational photometric variability are stable during at least ~ 6000 days (16 years, or some 950 nutational periods). The photometric data confirm presence of irregular variations of the precessional period. The nutation radial velocity variations have some delay relative to nutational photometric variations, by $0.^d6$ (see Figure 5).

This fact is in agreement with the model of a nodding accretion disk. The time delay of $0.^d6$ can be identified as a sum of the travel time of collimated matter from the center of the accretion disk to the optically emitting parts of relativistic jets and the time of tidal perturbation transmission from the outer parts of the accretion disk to the interior ones.

In Figure 6 the orbital V light curves of SS 433 are presented versus different phases of the precessional period.

One can observe considerable variability of the orbital light curve of SS 433 as a function of a phase of the precessional period. It is believed to be due to precession of the accretion disk surrounding the relativistic object. At the primary orbital minimum the optical star is in front of the precessing accretion disk. Significant variability of the light of SS 433 at the primary minimum of an eclipsing light curve with the phase of the precessional period (amplitude $0.^m5$) is confirmed. This fact suggests partial eclipses of the precessing accretion disk at all phases of the precessional period.

There is, however, some inconsistency between X-ray and optical results of investigations of SS 433. A high value of the X-ray eclipse width in SS 433 implies a small value of the mass ratio $q = m_x/m_v = 0.16$ (m_x, m_v are masses of the

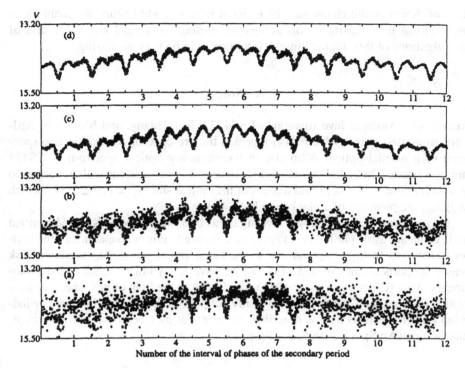

Figure 6. Shape of the orbital light curves as a function of precession phases: (a) the observed light curve; (b) the synthetic curve obtained by adding up all regular periodicities; (c) the synthetic curve obtained by adding up three periodicities: the orbital, precession and nutation components; (d) the synthetic curve obtained by adding up only two periodicities: the orbital and precession components. The synthetic curves were calculated for the times of observations. The orbital phases from −0.5 to +0.5 and the index number of an interval of the precession phase are plotted along the X axis.

relativistic object and the optical star, correspondingly). Such a small value of $q = 0.16$ corresponds to total eclipses of the precessing accretion disk at all phases of the precessional period because the orbital plane inclination is fixed, $i \simeq 79°$, and the optical star overfills its Roche Lobe. This fact is inconsistent with high-amplitude ($\sim 0.^m5$) variability of SS 433 at the primary minimum which depends on a phase of the precessional period. To solve this problem further X-ray and optical observations of SS 433 are needed (see the recent paper by Marshall *et al.*, 2002).

6. Conclusion

We have described the basic properties of the precessing accretion disk and relativistic jets in SS 433 binary system. In this complicated and enigmatic object a kind of deterministic chaos may be realized which would disturb the main clock precessional mechanism. In this connection further theoretical and observational investigations of this unique binary system seem to be very promising.

7. Notes

Recent spectroscopic investigations of SS433 (Gies, Huang, and McSwain, ApJ-Lett, submitted, astro-ph/0208044) revealed the presence of absorption line spectrum of a normal optical A7Ib star in the summary optical spectrum of SS433 binary system. The radial velocity shifts observed are consistent with a mass ratio $q = m_x/m_v = 0.57 \pm 0.11$ and masses of the optical star $m_v = (19 \pm 7) M_\odot$ which suggests the presence of a black hole in SS433.

Recent infrared spectroscopic investigations of SS433 (Fluchs, Koch-Miramond, and Abraham, astro-ph/0208432) revealed enhanced helium abundance of the matter transferred from the optical star to the supercritical precessing accretion disk which supports the model of SS433 as a massive X-ray binary at advanced evolutionary stage (Cherepashchuck, 1981a,b,c, 1988). In the light of new spectroscopic results of Gies *et al.*, large width of X-ray eclipses in SS433 may be due to wind-wind collision effects as was described in the paper by Cherepashchuk, Bychkov, and Seifina (Ap.Sp.Sci., 1995, v. 229, p. 33.

References

Antokhina, E. A., Cherepashchuk, A. M.: 1987, *Sov. Astron.*, **31**, 295.
Antokhina, E. A., Seifina, E. V., Cherepashchuk, A. M.: 1992, *Astron. Zh.*, **69**, 282.
Baykal, A., Anderson, S. F., Margon, B.: 1993, *Astron. J.*, **106**, 2359.
Bisikalo, D. V., Boyarchuk, A. A., Kuznetsov, O. A., Chechetkin, V. M.: 1999, *Astron. Zh.*, **76**, 672.
Borisov, N. V., Fabrika, S. N.: 1987, *Sov. Astron. Lett.*, **13**, 200.
Crampton, D., Cowley, A. P., Hutchings, J. B.: 1980, *Astrophys. J.*, **235**, L131.
Champton, D., Hutchings, J. B.: 1981, *Astrophys. J.*, **251**, 604.
Cherepashchuk, A. M.: 1981a, *MNRAS*, **194**, 761.
Cherepashchuk, A. M.: 1981b, *Pis'ma Astron. Zh.*, **7**, 201.
Cherepashchuk, A. M.: 1981c, *Pis'ma Astron. Zh.*, **7**, 726.
Cherepashchuk, A. M.: 1988, *Sov. Sci. Rev. Ap. Space Phys.*, R. A. Sunyaev (ed.), **7**, 1.
Cherepashchuk, A. M., Yarikov, S. E.: 1991, *Pis'ma Astron. Zh.*, **17**, 605.
Cherepashchuk, A. M., Katysheva, N. A., Khruzina, T. S., Shugarov, S. Yu.: 1996, *Highly Evolved Close Binary Stars*, Catalog, Cordon and Breach Publ., Amsterdam.
Clark, D. H., Murdin, P.: 1978, *Nature*, **276**, 45.
Collins, G. W. II.: 1985, *MNRAS*, **213**, 279.
D'Odorico, S., Oosterloo, T., Zwitter, T., Calvani, M.: 1991, *Nature*, **353**, 329.
Fabian, A. C., Rees, M. J.: 1979, *MNRAS*, **187**, 13P.
Fabrika, S. N., Bychkova, L. V.: 1990, *Astron. Aph.*, **240**, L5.
Gladyshev, S. A., Kurochkin, N. E., Novikov, I. D., Cherepashchuk, A. M.: 1979, *Astron. Circ.*, N1086.
Gladyshev, S. A. Goranskii, V. P., Cherepashchuk, A. M.: 1983, *Pis'ma Astron. Zh.*, **9**, 3.

Gladyshev, S. A., Goranskii, V. P., Cherepashchuk, A. M.: 1987, *Sov. Astron.*, **31**, 541.
Goranskii, V. P., Esipov, V. F., Cherepashchuk, A. M.: 1998a, *Astronomy Reports*, **42**, 209.
Goranskii, V. P., Esipov, V. F., Cherepashchuk, A. M.: 1998b, *Astronomy Reports*, **42**, 336.
Katz, J. I.: 1973, *Nature Phys. Sci.*, **246**, 87.
Katz, J. I.: 1980, *Astrophys. J.*, **236**, L127.
Katz, J. I.: 1997, *Astrophys. J.*, **478**, 527.
Katz, J. I., Anderson, S. F., Margon, B., Grandi, S. A.: 1982, *Astrophys. J.*, **260**, 780.
Kemp, J. C., Henson, J. D., Krauss, D. J., *et al.*: 1986, *Astrophys. J.*, **305**, 805.
Kotani, T., Kawai, N., Matsuoka, M., Brinkman, W.: 1996, *Publ. Astron. Soc. Japan*, **48**, 619.
Kotani, T.: 1998, *X-ray observations of a galactic jet system SS 433 with ASCA, ISAS Research Note*,
    N655, Univ. of Tokyo.
Maloney, P. R., Begelman, M. C., Pringle, J. E.: 1996, *Astrophys. J.*, **472**, 582.
Maloney, P. R., Begelman, M. C., Pringle, J. E.: 1997, *Astrophys. J.*, **491**, L43.
Mammano, A., Ciatti, F., Vittone, A.: 1980, *Astron. Aph.*, **85**, 14.
Margon, B., Ford, H. C., Katz, J. I., *et al.*: 1979, *Astrophys. J.*, **230**, L41.
Margon, B.: 1984, *Ann. Rev. Astron. Aph.*, **22**, 507.
Marshall, H. L., Canizares, C. R., Schulz, N. S.: 2002, *Astrophys. J.*, **564**, 941.
Martin, P. G., Rees, M. J.: 1979, *MNRAS*, **189**, 19P.
Matese, J. J., Whitmire, D. P.: 1982, *Astron. Aph.*, **106**, L9.
Milgrom, M.: 1979, *Astron. Aph.*, **76**, L13.
Newsom, G. H., Collins, G. W. II: 1982, *Astrophys. J.*, **262**, 714.
Panferov, A. A., Fabrika, S. N.: 1997, *Astron. Zh.*, **74**, 574.
Petterson, J. A.: 1975, *Astrophys. J.*, **201**, L61.
Pringle, J. E.: 1996, *MNRAS*, **281**, 357.
Roberts, W. J.: 1974, *Astrophys. J.*, **187**, 575.
Sarazin, C. L., Begelman, M. C., Hatchett, S. P.: 1980, *Astrophys. J.*, **238**, L129.
Shakura, N. I.: 1972, *Astron. Zh.*, **49**, 921.
Shakura, N. I., Sunyaev, R. A.: 1973, *Astron. Aph.*, **24**, 337.
Stewart, G. C., Watson, M. G., Matsuoka, M., *et al.*: 1987, *MNRAS*, **228**, 293.
Van den Heuvel, E. P. J., Ostriker, P., Petterson, J. A.: 1980, *Astrophys. J.*, **81**, L7.
Vermeulen, R. C., Murdin, P. G., van den Heuvel, E. P. J., *et al.*: 1993, *Astron. Aph.*, **270**, 204.

ORBITS AND INTEGRALS IN SELF-CONSISTENT SYSTEMS

G. CONTOPOULOS, N. VOGLIS and C. EFTHYMIOPOULOS

Academy of Athens, Research Center for Astronomy, 14 Anagnostopoulou Str., 10673 Athens, Greece

Abstract. We find the forms of the orbits in a self-consistent galactic model generated by a N-body simulation of the collapse of a protogalaxy. The model represents a stationary elliptical galaxy of type E5, which is approximately axisymmetric around its longest axis. The orbits are of three main types, box orbits (including box-like orbits), tube orbits and chaotic orbits. The box or box-like and tube orbits are represented by closed invariant curves on a Poincaré surface of section. The forms of the orbits and of the invariant curves can be explained by a third integral of motion I, that is given by the Giorgilli (1979) computer program. The nonresonant form of the third integral explains the box orbits, while a resonant form of this integral explains both the box orbits and the 1:1 tube orbits. The N-body model gives the distribution of velocities F, which is an exponential of the third integral.

1. Introduction

Most of the research work up to now on orbits and integrals of motion has been done in given potentials. These systems are in general not self-consistent, i.e. the potential is not necessarily produced by the distribution of the particles forming the system.

On the other hand, N-body systems of particles are by definition self-consistent if there are no external forces besides the attractions of the particles themselves. If the number N of bodies is large the system approximates a continuous potential.

The computer simulations of systems of N bodies can approximate astronomical objects like clusters and galaxies. A particularly important class of numerical simulations refers to the formation of galaxies. Let us consider a set of points following the expansion of the Universe, but with larger density in a certain domain, that can be called a protogalaxy. The points of this domain follow initially the general expansion of the Universe, but with smaller velocities, and later collapse to generate a galaxy (Figure 1).

In our numerical experiments we have considered several distributions of the density in such a protogalaxy and the final outcomes are different types of elliptical galaxies (Voglis, 1994; Efthymiopoulos and Voglis, 2001). In all cases the outcome of the collapse of the protogalaxy is a stationary model of a galaxy with very small oscillations, of the order of 1%.

A particular self-consistent model constructed by the collapse of a system of $N = 10^6$ bodies is shown in Figure 2. This is a triaxial system similar to an

Space Science Reviews **102**: 37–50, 2002.
© 2002 *Kluwer Academic Publishers. Printed in the Netherlands.*

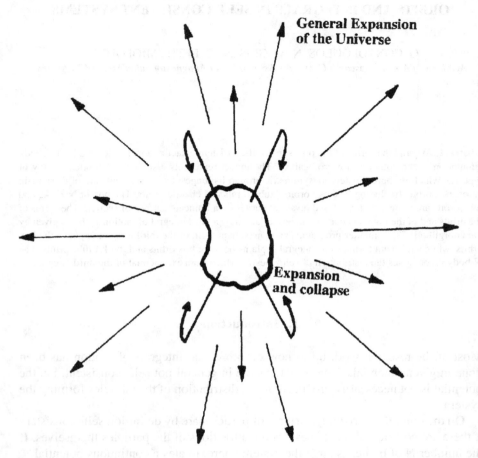

Figure 1. A schematic view of the collapse of a protogalaxy within the general expansion of the Universe.

elliptical galaxy of type E5. The longest axis is along the z-direction. In fact, the x, y and z axes are proportional to 1, 1.51, 2.04 (Figure 2(a)–(b)).

We can represent the potential of this system by using spherical Bessel functions and Legendre polynomials (Contopoulos *et al.*, 2000). This potential can be represented by an axisymmetric background

$$V(R, z) = \sum_{k=0}^{8} \sum_{l=0}^{8} g_{lk} R^{2k} z^{2l} \qquad (1)$$

where

$$R^2 = x^2 + y^2 \qquad (2)$$

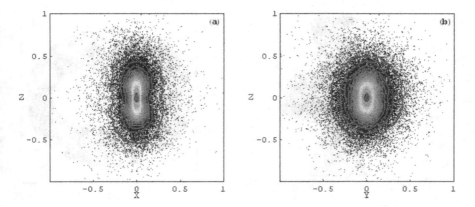

Figure 2. A particular self-consistent model of a galaxy produced by the collapse of a protogalaxy. Projections: (a) on the plane (x, z), and (b) on the plane (y, z).

plus a three-dimensional perturbation of the form $V_1(R, \phi, z)$, where ϕ is the azimuthal angle on the (x, y) plane. This perturbation is of the order of 10%. In a recent paper (Contopoulos *et al.*, 2002) we discuss the role of this perturbation, and also the role of the ellipticity of the galaxy, as regards the forms of the orbits. However, in the present paper we consider only the axisymmetric potential (1). We have taken enough terms (64) in order to fit the observed axisymmetric component of the potential of the system shown in Figure 2 with an accuracy better than 3%.

The Hamiltonian of the system (1) is

$$H \equiv \frac{1}{2}(\dot{R}^2 + \dot{z}^2 + \frac{L_z^2}{R^2}) + V = h \qquad (3)$$

where h is the energy of a particle (star) (first integral) and L_z its angular momentum (second integral).

In the following sections we consider only a nonrotating galactic system. In section 2 we study the orbits and their invariant curves. The various types of orbits can be explained by using appropriate third integrals in nonresonance and resonance cases (section 3). Finally we summarize our results and conclusions in section 4.

2. Orbits and Invariant Curves

There are two main types of orbits in a model galaxy of type (1), ordered and chaotic. The ordered orbits are divided into box orbits (including box-like orbits) (Figure 3(a)) and tube orbits of various types (Figure 3(b)–(c))[1]. Finally the chaotic orbits fill in an irregular way a large part of the space (R, z) (Figure 3(d)). The box

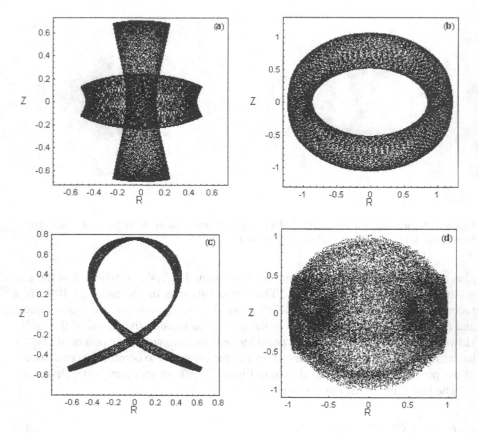

Figure 3. Various forms of orbits in the galactic model (1): (a) two different box orbits, (b) the main type of a tube orbit (type 1:1), (c) a tube orbit of type 3:2, and (d) a chaotic orbit.

orbits appear near the center $R = z = 0$ and near the axes $R = 0$ and $z = 0$ for all values of the energy h. The tube orbits and the chaotic orbits appear only for large values of the energy h.

The phase space of all these orbits is 4-dimensional $(R, z, \dot{R}, \dot{z})$. If we fix the energy h (and the angular momentum L_z) of the orbits we find a three-dimensional space $(R, z, \dot{R})$. If then we consider the intersections of the orbits by a Poincaré surface of section $z = 0$ we find the types of orbits in the plane $(R, \dot{R})$. In the cases of regular orbits (boxes, or tubes) the intersections of each orbit lie on smooth invariant curves (Figures 4(a) and 5(a)).

In the case of Figure 4 the value of the energy is $h = -1600000$, near the bottom of the potential well of the potential (1). The angular momentum L_z of the orbit is small ($L_z = 0.045$), but not zero. The central invariant point (periodic orbit of the map) on the surface of section $(R, \dot{R})$ has $R = 0.04$ and $\dot{R} = 0$. This point

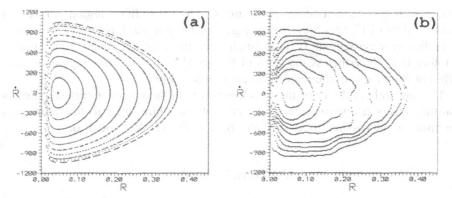

Figure 4. (a) Invariant curves of orbits on a Poincaré surface of section $z = 0$ $(R, \dot{R})$ for $h = -1600000$ and $L_z = 0.045$. These invariant curves represent box-like orbits. (b) Equidensities of the orbits on the plane $(R, \dot{R})$ (i.e. $z = 0$).

represents the intersections by the plane $z = 0$ of an orbit filling an axisymmetric surface in the 3-D configuration space (x, y, z). All the other orbits form closed invariant curves around this point, extending up to about $R = 0.37$. The existence of well defined invariant curves indicates that a 'third integral' is applicable in these cases.

In the case of Figure 5 the energy is larger ($h = -900000$) while $L_z = 0.04$. In this case we see three main types of orbits: (a) Box orbits around a stable periodic orbit of the map $(R, \dot{R} = 0)$. (b) Tube orbits of various types. The main tube orbits are of 1:1 type (Figure 3(b)) for relatively large values of R. However, there are also other types of tube orbits, forming islands of stability in Figure 5(a). (c) Chaotic orbits represented by scattered points in Figure 5(a). These orbits surround the main island of tube orbits and separate the box orbits from the main tube orbits. Note that the extent of Figure 5(a) reaches about $R = 0.97$, i.e. much larger distances than in Figure 4(a). In this case also most orbits form good invariant curves, therefore they are represented by a 'third integral' of motion.

We collect all the particles of the original N-body system with energies in small bins around the values of h and L_z, used in Figures 4(a) and 5(a). We run the orbits of these particles in the full triaxial self-consistent potential and we collect their points on the plane $(R, \dot{R})$ when $z = 0$ with $\dot{z} > 0$. In Figures 4(b) and 5(b) we mark the lines of equal density of these points on the plane $(R, \dot{R})$ (equidensities). These lines are not very accurate, because the number $N = 10^6$ bodies is divided into many values of the energy and angular momentum and into many cells that form a grid on each surface $(R, \dot{R})$. (In order to increase the number of bodies in Figures 4(b) and 5(b) we calculated the orbits with different z at a given time t until they intersect the plane $z = 0$).

By comparing Figure 4(b) with Figure 4(a), and Figure 5(b) with Figure 5(a) we find many similarities. In particular the equidensities of the box orbits are very similar with the invariant curves of Figures 4(a) and 5(a). In the region of tube and chaotic orbits of Figures 5(a)–(b) the agreement is not good, because the number of bodies in these regions is quite small, thus the accuracy of the equidensities is rather bad. Nevertheless the form of the equidensities in these regions is very different from those of the regions covered by box orbits.

This general similarity of the equidensities with the invariant curves indicates that the density is approximately a function of the 'third integral' that represents the invariant curves that fill most of the space in Figures 4(a) and 5(a).

3. Third Integrals

We now construct appropriate third integrals for the various types of regular orbits, namely the box and the 1:1 tube orbits.

Up to now the third integral (Contopoulos, 1960) has been used in simple dynamical systems, in cases where the potential V consists of quadratic terms, representing two harmonic oscillators, plus a few higher order terms. The first models represented a rough approximation of the potential of our galaxy around the Sun. Later we used some simple resonant forms of the third integral (Contopoulos, 1963) that are necessary in order to explain the forms of the tube orbits. In order to find higher order terms of the third integral we have constructed an appropriate computer program (Contopoulos and Moutsoulas, 1965) that gives such terms up to very high orders.

The most effective program of this form was developed by Giorgilli (1979). It can give formal integrals in systems of two or more degrees of freedom very fast.

In the present case, where the potential contains 64 terms the use of such an automatic method is absolutely necessary in order to get accurate results. In fact we

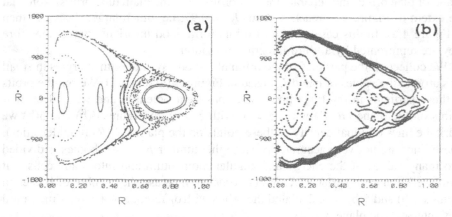

Figure 5. As in Figure 4 for $h = -900000$ and $L_z = 0.04$. The Poincaré surface of section contains: (1) closed invariant curves on the left (box-like orbits). (2) closed invariant curves forming an island on the right (tube orbits 1:1). (3) scattered points (chaotic orbits) and (4) higher order islands.

have used the Giorgilli program to calculate the third integral up to order (degree) 30.

The terms of the third integral up to order 4 are

$$
\begin{aligned}
I(R, z, \dot{R}, \dot{z}) = {} & 1.22560(R^2 + \dot{R}^2) - 0.86410R^4 + 0.51846\dot{R}^4 + \\
& 1.03692R^2\dot{R}^2 - 1.49732R^2\dot{z}^2 + 0.29895R^2z^2 + \\
& 1.4932z^2\dot{R}^2 - 2.07271Rz\dot{R}\dot{z} - 0.29895\dot{R}^2\dot{z}^2 + ...,
\end{aligned}
$$

(4)

in the nonresonant case, while they are

$$
\begin{aligned}
I(R, z, \dot{R}, \dot{z}) = {} & 1.22560(R^2 + \dot{R}^2 + z^2 + \dot{z}^2) - 0.86410R^4 + \\
& 0.51845\dot{R}^4 + 1.03692R^2\dot{R}^2 - 2.01768R^2z^2 + \\
& 0.81876R^2\dot{z}^2 - 0.059398z^2\dot{R}^2 + 1.51984Rz\dot{R}\dot{z} - \\
& 0.33755z^4 + 0.40506z^2\dot{z}^2 + ...,
\end{aligned}
$$

(5)

in the 1:1 resonant case. In the resonance case the third integral applies both to the box orbits and to the 1:1 tube orbits (but not to tube orbits of other types). In the above formulae (4) and (5) the coefficients of the third integral are normalized in units of the constant $g_{02} = 2.06 \times 10^6$, while the velocities are normalized in units of $\sqrt{g_{02}}$.

In Figure 6 we see a comparison of the invariant curves calculated numerically for $L_z = 0$ and of the level curves of the third integral. In the first case ($h = -1600000$) the agreement is perfect. In the second case ($h = -900000$) the agreement is fairly good for the box orbits, but less good for the 1:1 tube orbits. In fact, the theoretical level curves form islands of type 1:1, but these are displaced

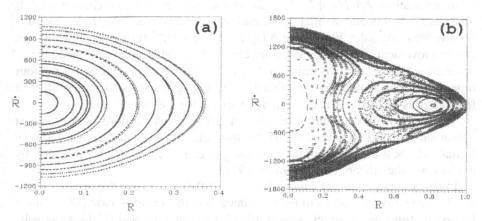

Figure 6. Invariant curves (——) and theoretical level curves of the third integral (dotted curves in (a) and squares in (b)) for $L_z = 0$ and :(a) $h = -1600000$, (b) $h = -900000$.

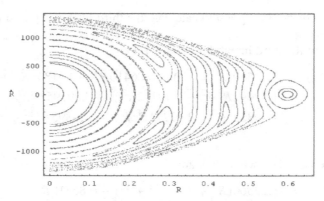

Figure 7. Invariant curves of box orbits and tube orbits of the type 3:2 of Figure 3(c) (upper island around $R = 0.25$) and inverted orbits of type 3c (lower island).

with respect to the real islands. Furthermore, the theoretical islands cover also the chaotic domain, where the third integral is not applicable.

Other forms of the third integral appropriate for other resonances were also computed, but they are of small importance because they are applicable only around some small islands of stability. Such is the case of the resonant orbits of type 3:2 (Figure 3(c)) that gives two particular islands on a Poincaré surface of section (Figure 7).

The most important case of the third integral is the nonresonant one, that represents the box orbits. In the case of Figure 4 (where $h = -1600000$) we calculated the third integral along several orbits to find their constancy. In every case we truncate the third integral after a certain order. In Figure 8 we see that the third integral is better conserved as the order of truncation increases up to a certain degree. In particular if we truncate the third integral at order 6 (I_6) we find that the relative variations $\Delta I_6/I_6$ of I_6 along an orbit are of order 6×10^{-2} (Figure 8(a)). If we truncate the integral at order 10 we find relative variations $\Delta I_{10}/I_{10} \approx 2 \times 10^{-3}$ (Figure 8(b)). At order 20 we have $\Delta I_{20}/I_{20} \approx 1.3 \times 10^{-5}$ (Figure 8(c)). However, this improvement of accuracy does not continue at higher orders. In particular at order 30 we have a relative variation $\Delta I_{30}/I_{30} \approx 5 \times 10^{-5}$ (Figure 8(d)), i.e. four times larger than at order 20. Note that the scale of Figures 8(c)–(d) is very different from Figures 8(a)–(b), so that in Figures 8(c)–(d) we see more details.

Thus there is a maximum order (around 20) at which the third integral is best conserved. This is due to the fact that the third integral is not convergent, but only formal. The best truncation of the third integral can be calculated theoretically by the Nekhoroshev theory (Nekhoroshev, 1977).

In the resonance case of Figure 5 ($h = -900000$) we have the results of Figure 9. The relative variations of the third integral in this resonant case $\Delta I_i/I_i$ along box orbits truncated at orders $i = 8, 10, 14$ are $0.34, 0.06$ and 0.18 respectively. The best truncation in this case is at order 10 (I_{10}).

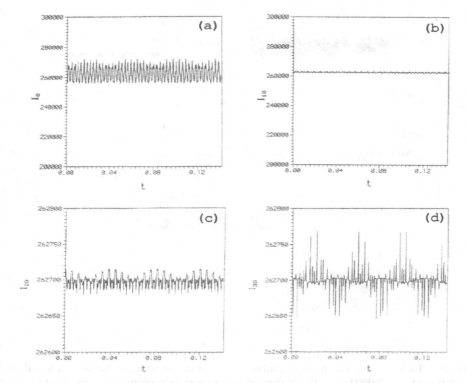

Figure 8. The variations of the third integral along an orbit with initial conditions $R = 0.2$, $\dot{R} = 0$ and $h = -1600000$, $L_z = 0$. The third integral is truncated at orders :(a) 6, (b) 10, (c) 20 and (d) 30.

The third integral is now used in finding a relation between the density

$$F = \frac{dN}{dh\, dL_z\, dR\, d\dot{R}} \qquad (6)$$

of particles in the intervals $(h, h + dh)$, $(L_z, L_z + dL_z)$ of energy and angular momentum and in small squares $(dR, d\dot{R})$ on the surface of section. For fixed intervals of h and L_z, the density F is a function of R and $\dot{R}$:

$$F = F(R, \dot{R}) \qquad (7)$$

Figure 10 gives the density F: (a) as a function of R for $\dot{R} = 0$, and (b) as a function of $\dot{R}$, for a representative value of $R = 0.18$, for $h = -1600000$ and $L_z = 0.045$.

The maximum of F for $\dot{R} = 0$ (Figure 10(a)) is not at $R = 0$, because $L_z \neq 0$. On the other hand the maximum F for fixed R is near $\dot{R} = 0$ (Figure 10(b)). The function $F(\dot{R})$ in Figure 10(b) has variations of order 15%. This is to be expected because the number of points $N = 10^6$ is divided among various energies and angular momenta and then among 20×20 squares $(\Delta R, \Delta \dot{R})$ on the $(R, \dot{R})$ plane.

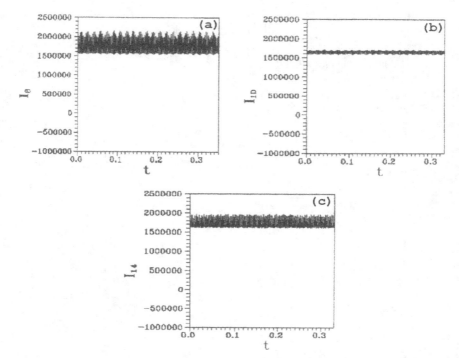

Figure 9. The variations of the resonant third integral along an orbit with initial conditions $R = 0.79$, $\dot{R} = 0$ and $h = -900000$, $L_z = 0$. The third integral is truncated at orders :(a) 8, (b) 10, (c) 14.

The total number of points on a given surface of section is about 6000, and about 200 squares are occupied on the plane $(R, \dot{R})$. Thus an average number of points in every square is $N_s = 30$ with deviations of order $\sqrt{N_s} = 5.5$. This is of the same order as the variations in Figure 10(b). These variations produce also the non smooth equidensity curves of Figure 4(b).

If now we compare the logarithm of the density F with the value of the third integral I, we find a roughly linear relation (Figure 11). This shows that F is given by an exponential formula

$$F = ae^{-bI} \tag{8}$$

The best fit is achieved with $a = 94.6$ and $b = 3.72$, using the values of I truncated at order 20.

In the resonant case of Figure 5 the function F has large values for the box orbits near the center (Figure 12(a)), while the density F drops almost to zero near the end of the region of the box orbits at $R \approx 0.55$ (Figures 5(a) and 12(a)). Then there are two small peaks beyond $R \approx 0.55$ (Figure 12(a)) and a minimum between them at about $R = 0.8$. The minimum corresponds to the center of the island of

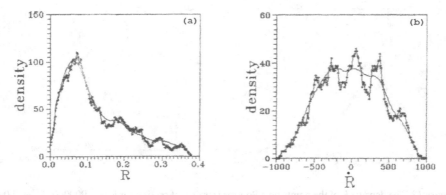

Figure 10. The density F of points on a Poincaré surface of section $(R, \dot{R})$: (a) as a function of R for $\dot{R} = 0$, and (b) as a function of $\dot{R}$ for constant $R = 0.18$, in the case $h = -1600000$ and $L_z = 0.045$ of Figure 4. The smooth lines are given by an interpolation formula.

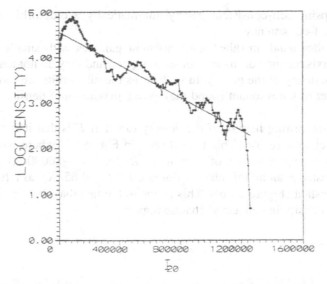

Figure 11. The logarithm of the density F as a function of the third integral I_{20} (i.e. truncated at order 20).

tube orbits of Figure 5a. This minimum is seen even better in Figure 12(b), where the density F is given as a function of $\dot{R}$ for $R = 0.77$.

The fact that the density of the tube orbits is minimum at the center of the island seems at first strange. However, one notices that the main tube orbits are elliptical, elongated perpendicularly to the z-axis of the galaxy (Figure 2(b)). Therefore one cannot have many orbits of this type, because they tend to destroy the bar. If we had many stars along such orbits the galaxy would not be self-consistent. The evolution

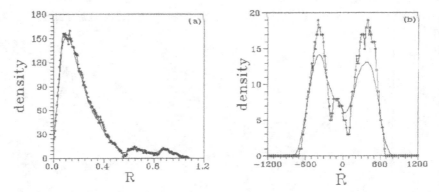

Figure 12. The density F of the points on a Poincaré surface of section for $h = -900000$, $L_z = 0.04$: (a) $F(R)$ for $\dot{R} = 0$, and (b) $F(\dot{R})$ for $R = 0.77$. The smooth lines are given by an interpolation formula.

of the collapsing self-consistent galaxy automatically avoids orbits that tend to destroy its self-consistency.

On the other hand, in other self-consistent galaxies with smaller elongation along the z-axis the tube orbits are not so elongated and they do not tend to destroy the self-consistency of the galaxy. In such cases the tube orbits are more abundant and the center of the resonant island may give a maximum of density, instead of a minimum.

Another interesting feature of the density function F is that it is roughly constant in the chaotic region. This is best seen in Figure 13, where we see a large chaotic domain on the surface of section $(R, \dot{R})$ for $h = -800000$ (Figure 13(a)). Then the density F along this chaotic domain $F(R = 0.65, \dot{R})$, as a function of $\dot{R}$ is almost constant (Figure 13(b)). This is due to the fact that chaotic orbits tend to equalize the density in connected chaotic regions.

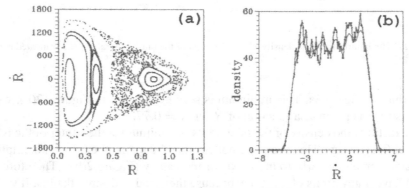

Figure 13. (a) A Poincaré surface of section $(R, \dot{R})$ for $h = -800000$ and $L_z = 0.04$. (b) The density $F(\dot{R})$ for $R = 0.65$. Along the chaotic domain, F is approximately constant density.

4. Conclusions

(1) The numerical N-body calculations give by definition self- consistent models of galaxies, i.e. they give not only the density distribution but also the distribution of velocities. A given set of initial conditions gives after an initial phase of cosmological expansion and collapse, a final distribution which is almost completely stationary.

(2) In our case we constructed a triaxial model of an E5 galaxy, with the longest axis along the z-direction and we fitted the axisymmetric background around the longest axis z by using an appropriate potential $V(R, z)$. In this potential there are three types of orbits: (a) box or box-like orbits, (b) tube orbits of various resonance types, and (c) chaotic orbits. Similar results were found in other models also. Thus, in general a galaxy contains both ordered and chaotic orbits. The ordered orbits form closed invariant curves on a Poincaré surface of section, while the chaotic orbits are represented by scattered points.

(3) The forms of the invariant curves can be found by means of a third integral of motion. This integral has a nonresonant form and many resonant forms, each form applicable to a single resonance. The third integral is not applicable to the chaotic orbits.

(4) The constancy of the third integral was checked for ordered orbits, both in the nonresonant case and in a particular resonant case, applicable to the elliptical (1:1) tube orbits. In every case we used truncated forms of the third integral. The variations of the third integral along an orbit decrease as the order of truncation increases up to a certain order. But if the truncation is done at higher orders the variations of the third integral are larger. This behaviour is consistent with the Nekhoroshev theorem.

(5) The distribution of the velocities is given by a distribution function F, which is well represented in our examples, by an exponential of the form $F = a \exp(-bI)$, at least for the box-like orbits.

(6) The distribution function has a minimum for tube orbits of type 1:1. This is explained because such orbits are perpendicular to the z-axis, therefore if they were abundant, they would tend to destroy the self-consistency of the model.

(7) The distribution function is almost constant for chaotic orbits in the same connected chaotic domain. This is explained because the chaotic orbits tend to populate equally the various parts of a chaotic domain.

5. Note

1. The terms box and tube orbits were introduced by Torgard and Ollongren (1960). The tube orbits surround stable resonant periodic orbits of various types. The theoretical explanation of the forms of the box and tube orbits was given by using nonresonant and resonant forms of the third integral (Contopoulos 1963,1965, Contopoulos and Moutsoulas 1965). The plane orbits shown in Figure 3(a) become 3-D box orbits if an oscillation along the x-axis is superimposed,

provided that $L_z = 0$. If $L_z \neq 0$ these orbits acquire a hole along the z-axis and they are known as 'inner long axis tube' (see e.g. de Zeeuw 1985). However, if L_z is small, as in the cases considered in the present paper, the overall form of these orbits is very similar to boxes (despite the hole at their center) and very different from the other types of tube orbits. Thus we call these orbits 'box-like' in order to distinguish them clearly from the other types of orbits, and thus simplify the classification.

Acknowledgements

This research was supported in part by the Academy of Athens (research program 200/493).

References

Contopoulos, G.: 1960, *Z. Astrophysik* **49**, 273.
Contopoulos, G.: 1963, *Astron. J.* **68**, 763.
Contopoulos, G.: 1965, *Astron. J.* **70**, 526.
Contopoulos, G. and Moutsoulas, M.: 1965, *Astron. J.* **70**, 817.
Contopoulos, G., Efthymiopoulos, C., and Voglis, N.: 2000, *Cel. Mech. Dyn. Astron.* **78**, 243.
Contopoulos, G., Voglis, N. and, Kalapotharakos, C.: 2002, *Cel. Mech. Dyn. Astron.* (in press).
de Zeeuw, T.: 1985, *Mon. Not. R. Astr. Soc.* **216**, 273.
Torgard, I. and Ollongren, A.: 1960, Nuffic International Summer Course in Science, Part X.
Efthymiopoulos, C. and Voglis, N.: 2001, *Astron. Astrophys.* **378**, 679.
Giorgilli, A.: 1979, *Computer Phys. Comm.* **16**, 331.
Nekhorosev, N.N.: 1977, *Russ. Math. Surv.* **32**, 1.
Voglis, N.: 1994, *Mon. Not. R. Astr. Soc.* **267**, 379.

OBSERVATIONAL MANIFESTATION OF CHAOS IN THE GASEOUS DISK OF THE GRAND DESIGN SPIRAL GALAXY NGC 3631

A.M. FRIDMAN, O.V. KHORUZHII and E.V. POLYACHENKO

Institute of Astronomy of the Russian Academy of Sciences, Pyatnitskaya st. 48, Moscow, Russia

Abstract. The main goal of the paper is to demonstrate the presence of chaotic trajectories in the gaseous disk of a real spiral galaxy. As an example we have chosen NGC 3631. First, we show the stationarity of the 3-D velocity field restored from the observed line-of-sight velocity field of the gaseous disk. That allows to analyse behaviour of the trajectories of the fluid particles (gas clouds) in the disk, calculating the corresponding observed streamlines. We estimate the Lyapunov characteristic numbers using their independence of the metrics and show the existence of chaotic trajectories outside the vortices which are present in the velocity field, and in the vicinity of the saddle point. Related spectra of the stretching numbers for some trajectories are also calculated.

1. Introduction

NGC 3631 is a rather bright non-barred galaxy (of type SAc) with a well-defined spiral structure (Figure 1), at a distance of 15.4 Mpc, as estimated from its recession velocity using the Hubble constant of 75 km s^{-1} Mpc^{-1}, which gives the angular scale of 75 pc per arcsec. Interestingly, the galaxy has been included in Arp's (1966) atlas of peculiar galaxies, thanks to its 'straight arms', and 'absorption tube crossing from inside to outside of southern arm'. These features can be recognized in the *R*-image shown in Figure 1, but the galaxy as a whole looks to us rather normal. The atomic hydrogen distribution has been described by Knapen (1997 and references therein to the earlier work), and the ionized hydrogen has been studied through emission in the Hα line by, among others, Boeshaar & Hodge (1977), Hodge (1982) and Rozas, Beckman & Knapen (1996), Fridman *et al.* (1998), (2001a).

The galaxy optical axial ratio is close to unity: according to the RC3 catalogue (Vaucouleurs *et al.*, 1991), log a/b = 0.02 ± 0.07, so this galaxy is observed nearly face-on. Such an orientation is very favourable for studying the gas motions perpendicular to the plane of the galaxy, which was the main topic of a paper by Fridman *et al.* (1998).

In this paper it was shown that the non-circular gas motions in NGC 3631 have regular character, and that they are related to the observed two-armed spiral structure, which has a wave nature. The vertical (that is, perpendicular to the plane of the disk) component of the gas motions as revealed by a Fourier analysis method (Fridman *et al.*, 1997), was also found to be induced by the spiral density wave. The inclination angle of the disk of NGC 3631 was found to be about 17°, which

Space Science Reviews **102**: 51–72, 2002.
© 2002 *Kluwer Academic Publishers. Printed in the Netherlands.*

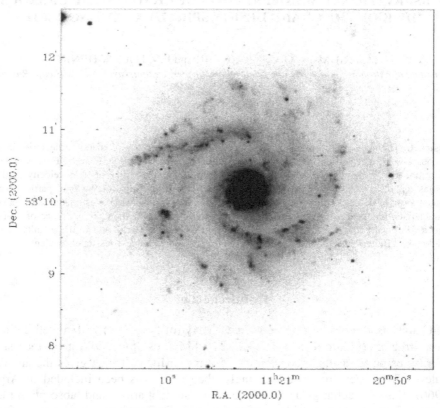

Figure 1. *R*-band image of NGC 3631 as obtained from the ING archive.

enables, using the same observational data, to restore the vector velocity field in the plane of this galaxy (Fridman *et al.* 2001a).

The analysis of the velocity field of NGC 3631 gaseous disk was based in Fridman *et al.* (2001a) on two types of line-of-sight velocity data, well complementing each other: in the radio HI and optical Hα lines (see below). The HI observations used for this study were obtained by Knapen (1997) with the Westerbork Synthesis Radio Telescope, and the Hα observations were carried out at the Special Astrophysical Observatory (SAO) with its 6-m reflector equipped with an F/2.4 focal reducer and a scanning Fabry–Perot interferometer.

The wave nature of NGC 3631 two-armed spiral structure was confirmed and its corotation radius was determined (about 42″ or 3.2 kpc). It was found that the projection of the restored three-dimensional vector velocity field of the gas in the plane of the galaxy, and in a reference frame corotating with the spiral pattern, reveals the presence of two anticyclonic vortices near corotation. The existence of cyclonic vortices in NGC 3631 was revealed, apart from the existence of an-

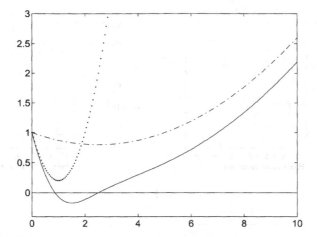

Figure 2. Dimensionless squared eigen frequency (ω^2/κ^2) in the gaseous disk alone (dashed line), the stellar disk alone (dotted line) and the combination of both disks (solid line) in a grand design spiral galaxy as a function of dimensionless wave vector ($k/[\pi G\sigma_*/c_*^2]$).

ticyclones mentioned above. Such cyclonic vortices are a consequence of a high amplitude of the density wave in this galaxy (Fridman *et al.* 2001a).

The present paper is devoted to the further analysis of NGC 3631 velocity field. Particularly, we discuss the possibility of existence of chaotic trajectories in this galaxy. First, we give a review of the dynamical characteristics of the galaxy determined up to now. Then, we calculate the maximum Lyapunov characteristic number (LCN) for different families of trajectories of the gas in the galactic disk. The LCN turs out positive, that argues in favour of a chaotic character of the trajectories.

2. Gravitational Instability in Grand-Design Spiral Galaxies

Figure 2 demonstrates the simplest model explaining the reason of the visual appearance of grand-design spiral galaxies. In a grand design spiral galaxy, the stellar disk in the absence of the gaseous disk is marginally stable, while the gaseous disk in the absence of the stellar disk is much more stable. As a result of the gravitational interaction between the two disks, an instability appears in a narrow range of the azimuthal wave numbers only (Figure 2).

The latter circumstance allows a simplified description of the large-scale velocity field in the grand-design spiral galaxies in an approximation of a wave with a particular azimuthal number (Fridman *et al.*, 1997, 2001a, 2001b). The approximation is confirmed by the predominance of the second Fourier harmonic in the spectrum of the surface density (Figure 3) and of the first, second and third Fourier

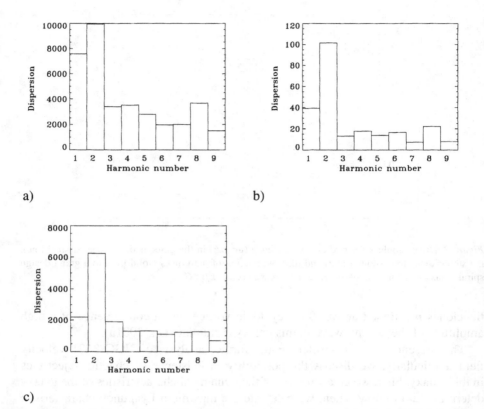

a)

b)

c)

Figure 3. Contribution of the individual Fourier harmonics to the dispersion in the model of axially symmetrical distribution in NGC 3631 (for details see the text), as derived from (a) the Hα image (6-m telescope SAO RAS), (b) the R-band image (ING archive), and (c) the 21 cm map (Knapen, 1997).

harmonics in the spectrum of the line-of-sight velocity field[1] (Figure 4) (Fridman *et al.*, 1998, 2001a).

In Figure 3 we can see a hystogram of the Fourier coefficient squares, $\bar{a}_m^2$, describing the expansion of the surface brightness of NGC 3631 in terms of the azimuth.

It is easy to verify that the dispersion D in the model representation of the surface brightness I as nonaxisymmetrical is $D \propto \sum_{m=1} \bar{a}_m^2$, where vinculum denotes radial averaging. Indeed, according to the definition $D = \sum_{i=1}^{N} [I_i - I_{sym}(r_i)]^2/N$, where N is the number of the observational points, I_i is the brightness measured in the i-th point, $I_{sym}(r_i)$ is the model axisymmetric brightness at radius r_i. The observed disk region can be split into K galactocentric rings, so that the dispersion

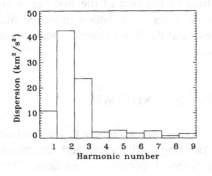
a)

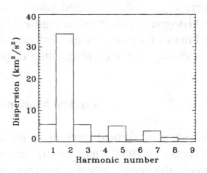
b)

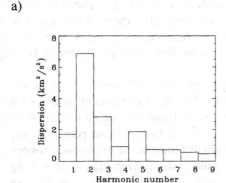
c)

Figure 4. Contribution of individual Fourier harmonics of the line-of-sight velocity field of NGC 3631 to the dispersion (for details see the text) in the model of pure circular motion in the region $r <$ 80″, as derived from (a) the original Hα line-of-sight velocity field, (b) Hα velocity field smoothed to a resolution comparable to that of the 21 cm data, (c) the 21 cm velocity field.

can be presented in the form

$$D = \frac{1}{N} \sum_{k=1}^{K} \sum_{l=1}^{L_k} \left[\sum_{m=1}^{M} a_m(r_k) \cos(m\varphi_l - \Phi_m) \right]^2 \propto \sum \bar{a}_m^2, \quad \sum_{k=1}^{K} L_k = N,$$

where L_k is the number of observational points in k-th ring and a_m is the amplitude of m-th azimuthal Fourier harmonics of I distribution in the ring.

In a similar manner one can show that the dispersion of the approximation of the line-of-sight velocity field in the model of pure circular orbits is

$$D = \frac{1}{N} \sum_{i=1}^{N} [V_i - V_{rot}(r_i)]^2 \propto \bar{a}_0^2 + \bar{a}_{\sin \varphi}^2 + \sum_{m=2} \bar{a}_m^2$$

where $\bar{a}_{\sin\varphi}^2$ is the averaged squared amplitude of the sine of the first harmonic. So histograms presented in Figures 3 and 4 show the contribution of the particular Fourier harmonic of the surface brightness and the line-of-sight velocity field, respectively, in the corresponding dispersion.

3. Vertical Motions of the Gas in NGC 3631

Two types of vertical motions in galactic disks are known (see e.g. Fridman and Polyachenko 1984): 'membranic' motions ('bending oscillations') and vertical motions in the density wave, which we call the 'vertical spiral-wave' motions. These two types of motions have different properties (Fridman and Polyachenko, 1984), in terms of dispertion as well as symmetries with regard to the $z = 0$ plane. The velocities of the membrane oscillations are even functions of z, $v_z(z) = v_z(-z)$, while the velocities of the vertical spiral-wave motions are odd functions of z, $v_z(z) = -v_z(-z)$.

Until now, the main method of studying membrane oscillations has been, essentially, observing warps in the galactic disks. The most convenient galaxies for such observations are those viewed nearly edge-on. However, such a method does not make it possible to say anything about the velocities of these perturbations, i.e., about their dynamical properties. It is virtually impossible to choose between different models and to determine the origin of the observed warping pattern having to base the conclusions on purely morphological information – just as in the case of the galactic spiral structures.

In the case of the density waves, the studies of the vertical motions are impeded by their relative smallness (as a consequence of the smallness of the parameter kh, where k is the absolute value of the wave vector, and h is the disk half-thickness). Therefore, their contribution to the observed line-of-sight velocities in galaxies is relatively important only for the galaxies that are viewed nearly face-on, so that the contribution of the motions in the disk plane is suppressed (proportional to $\sin i$, where i is the inclination angle of the galaxy). Thus, studies of both types of the vertical motions require analysis of the velocity fields in the galaxies that are viewed face-on.

Another important circumstance is associated with the fact that when determining the vertical (z) component of the velocity of some part of a disk viewed face-on, we determine the 'mean' velocity of this section over the entire 'visible' thickness of the disk.

If the optical depth of the disk at the wavelength considered is modest (as is true for the 21-cm line), the contributions of the nearest and farthest halves of the disk to the observed line-of-sight velocity are comparable. In this case, the observed amplitude of the spiral-wave motions should be small, since the motion on different sides of the $z = 0$ plane display mirror symmetry (if the optical depth is zero, the mean velocity is zero – the contributions of the two halves precisely compensate each other). In the case of a large disk opacity, as in the Hα line, the amplitude of the spiral-wave motions is substantially greater, since we effectively measure the velocity of the nearby outer boundary of the disk, where this velocity is maximum (the vertical velocity of the gas in the density wave grows with z). The velocities of the vertical motions in membrane oscillations are virtually constant throughout the disk thickness (due to the smallness of the parameter of kh); therefore, the ratio of the velocity amplitudes measured at optical and radio wavelengths should not depend significantly on the difference in the optical depths in these two observing bands. Comparing the velocity fields for a galaxy viewed face-on, which were found using optical and radio data, makes it possible to draw conclusions not only about the velocity distribution for vertical motions, but also about the origin of these motions.

For the galaxy NGC 3631, the results presented in Figure 4 are consistent with the hypothesis that the second Fourier harmonic in the line-of-sight velocity field spectrum of NGC 3631 is associated with the vertical gas motions. Indeed, in the case of the optical data, only the second harmonic has a significant amplitude, which can easily be understood if precisely this harmonic is associated with the vertical motions, while the effect of projection decreases the contributions of the radial and azimuthal velocities, which are proportional to $\sin i$ (in the case of NGC 3631, by more than a factor of three: $\sin i = \sin 16.5° \approx 0.28$).

At the same time, the amplitude of the second harmonic for the radio data is substantially lower than for the optical data[2]. The latter argues in favour of the fact that in NGC 3631 we observe the vertical spiral-wave motions rather than the bending oscilations. (For details see Fridman *et al.*, 1998, 2001a.)

4. Vector Velocity Field in the Disk Plane

Using the relations between the phases of different characteristics in the density wave we can draw a schematic picture for the velocity field with vortices superimposed on the surface density distribution (see Figure 5) (Fridman *et al.*, 1999).

In Figure 5 b one can see the presence of cyclones and anticyclones in the residual velocity field (that corresponds to a reference frame locally corotating with the disk at each radius, $V_{circ}(r) = 0$, where $V_{circ} \equiv V_{rot} - V_{r.f.}$, V_{rot} and $V_{r.f.}$ are the disk and the reference frame rotation velocities, respectively. This fact by itself does not depend on the amplitude of the spiral density wave, although the sizes of the vortices depend on this amplitude.

Lindblad and Langebartel (1953) were the first to calculate the field of displacements of the stars in the gravitational potential of a bar. The form of the star displacement field resembles the system of two cyclones and two anticyclones, which is similar to our schematic Figure 5 b. Referring to Lindblad and Langebartel (1953) Lynden-Bell (1996) has noted the existence of cyclonic and anticyclonic

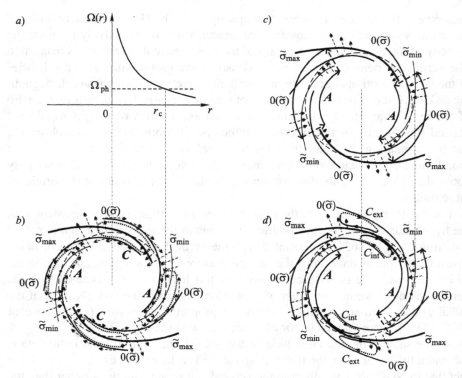

Figure 5. A picture for the velocity field with vortices in different reference frames. The vectors of the unperturbed rotation velocity field are shown by solid arrows, and the radial and azimuthal components of the residual velocity field – by dashed arrows. Solid curves of different thickness – the thickest, the thinnest, and intermediate – trace the azimuthal locations of the maxima, minima and zero values of the perturbed surface density defined at every radius and denoted by $\tilde{\sigma}_{max}$, $\tilde{\sigma}_{min}$, and $0(\tilde{\sigma})$ respectively. A and C denote anticyclones and cyclones correspondingly. a) In the laboratory reference frame the angular velocities of the disk, $\Omega(r)$, and of the spiral pattern, Ω_{ph}. The dashed line represents the corotation circle. b) The residual velocity field, which is the result of subtraction of the rotation velocity from the full velocity field. Dotted lines show the boundaries of the vortices. c) In the reference frame rotating with the angular velocity Ω_{ph}. The case when the gradient of the perturbed azimuthal velocity is less than the gradient of the circular velocity, $|\partial \tilde{V}_\varphi/\partial r| < |dV_{circ}/dr|$ $\equiv |d(V_{rot} - \Omega_{ph}r)/dr|$. One can see the existence of two anticyclones only. d) The same reference frame under the opposite condition $|\partial \tilde{V}_\varphi/\partial r| > |dV_{circ}/dr|$. One can see four cyclones along with two anticyclones.

trajectories in Figure 9 of their work. A similar picture of four vortices in the velocity field of a gaseous disk – two cyclones on the bar and two anticyclones between the spiral arms – was obtained in a recent paper by England *et al.* (2000).

In the reference frame rotating with the angular velocity Ω_{ph} of the two-armed spiral pattern, $V_{r.f.} = r\Omega_{ph}$, the picture of the vortices (see Figure 5 c, d) differs qualitatively from the previous one Figure 5 b. In the case when the gradient of the perturbed azimuthal velocity is less than the gradient of the circular velocity

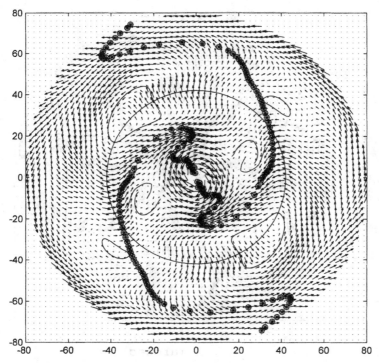

Figure 6. Restored velocity field of NGC 3631 in the plane of the disk in the reference frame rotating with the pattern speed. Asterisks show the locations of the maxima of the second Fourier harmonic of the R-band brightness map of the galaxy. The thin circle marks the position of the corotation. Solid lines demonstrate the vortex separatrices or nearly closed streamlines in the absence of a separatrix.

$(|\partial \tilde{V}_\phi/\partial r| < |dV_{circ}/dr| \equiv |d(V_{rot} - \Omega_{ph} r)/dr|)$, i.e. the anticyclonic shear dominates over the cyclonic one, the appearance of cyclones is impossible – we can see only two anticyclones with their centers near the corotation circle between two spiral arms (Figure 5 c). In the opposite case when the gradient of the perturbed azimuthal velocity dominates over the circular velocity gradient $(|\partial \tilde{V}_\phi/\partial r| > |dV_{circ}/dr|)$, the location of the cyclones is determined by the place of this domination. Cyclones can survive almost in the same places as shown in Figure 5 b, i.e. on the corotation circle, but their sizes should be smaller than those of the anticyclones. There are also two other variants when the cyclone centers are located either inside or outside the corotation circle along the zero lines of the radial velocity. Finally, both these last variants can be realized simultaneously. In this case four cyclones should exist in the velocity field as is shown in Figure 5 d. This is just the case which is observed in NGC 3631 that can be seen in Figure 6 (for details see Fridman *et al.* 2001a).

It is interesting to note, that both anticyclones and outer cyclones have closed separatrices whereas the inner cyclones have not. At first glance, this fact does

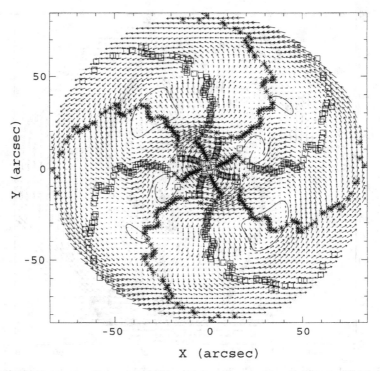

Figure 7. Lines of constant phases of the vertical motions superimposed on the velocity field of NGC 3631 in the plane of the disk in the reference frame rotating with the pattern speed. Squares mark the maxima of the absolute values of the vertical velocity of gas at each radius. Asterisks show the locations of the zeros of the vertical velocities. Note that the centers of the closed vortices lie close to the zeros, while the centers of the unclosed cyclones lie between the maxima and the zeros of the vertical velocity.

not agree with the hypothesis about a quasi-stationary character of the density wave in the galaxy. To overcome this apparent contradiction we should take into account that the real velocity field in the galactic disk is three-dimensional, but not two-dimensional. In Figure 7, the lines of constant phases of the second Fourier harmonic of the line-of-sight velocity field of NGC 3631 are superimposed on the vector velocity field in the disk plane shown in Figure 6. As shown in Sec. 3 (see also Fridman *et al.*, 1997, 1998, 2001a,b) the second harmonic of the line-of-sight velocity field characterizes the behaviour of the vertical motions in the density wave. It is clear from Figure 7 that the centers of the anticyclones and cyclones with closed separatrices lie close to the zeros of the vertical motions. On the contrary, the centers of the unclosed cyclones lie between the maxima and the zeros of the vertical motions. Thus the pattern presented in Figure 6 is in accordance with the quasi-stationary character of the density wave, but represents only two components of full three-component velocity field of the gas.

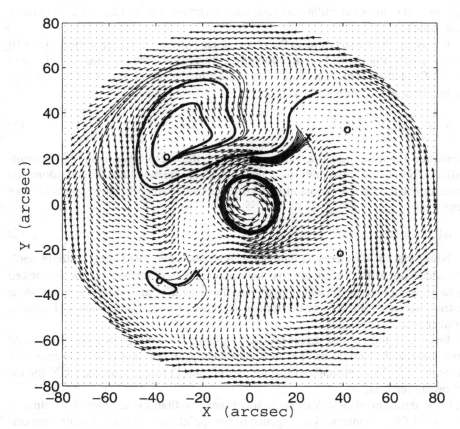

Figure 8. The two dimensional velocity field with some trajectories superimposed. Crosses correspond to the saddle points and circles to the centers of the vortices. Thick solid lines display reference
trajectories (for the definition see below) near the cyclone and anticyclone separatrices. Thin lines
show some trajectories that diverge from the reference trajectories. Also, a set of trajectories is shown
near a saddle point.

5. Dynamical Characteristics of Some Families of Gas Trajectories Outside the Vortices

We use the restored velocity field in the disk plane to investigate the behaviour
of some trajectories of fluid particles (clouds) in the vicinity of the cyclones and
anticyclones, as well as in a neighbourhood of the saddle point (Figure 8). In doing
so we use the coincidence between trajectories and stream lines for a stationary
velocity field. We focus our attention on the fact that in all the regions mentioned
above the trajectories that are initially close, become widely separated. To check
if such a property is a consequence of a chaotic behaviour of the system in these

regions first of all we calculate the Lyapunov characteristic numbers (LCN) of these trajectories.

Let us consider two points in the phase space with an initial separation $\vec{d}(t = 0)$ $= \vec{d}_0 = \vec{d}_1(0)$ (Figure 9). Moving along their trajectories, these points are separated by $\vec{d}_1(T)$ at the moment $t = T$. According to the definition of LCN (see e.g. Lichtenberg & Lieberman, 1983):

$$\lambda = \lim_{t \to \infty, \, d_0 \to 0} \frac{1}{t} \ln \frac{d_1(t)}{d_0} . \tag{1}$$

(hereafter, the absolute value is assumed when the vector over a variable is omitted). As follows from the definition, the LCN characterizes the averaged (along the trajectory) local rate of exponential divergence of two initially close points at large time scales:

$$d(t) \sim d_0 e^{\lambda t} . \tag{2}$$

Note, that in the theory (Lichtenberg & Lieberman, 1983) one restricts one's consideration to the linear approximation, which is described by the linearized dynamical equations. In addition to this theoretical restrictions, there are some restrictions that follow from the typical features of our observations and possess a rather general character.

Indeed, in the case of a real physical system, the concept of *close points* and *large time scales* should be specified. Particularly, in our case of the gaseous disk of NGC 3631, it is difficult to use the definition (1) to evaluate the LCN for the following reasons:

(1) the duration of observations is much smaller than the characteristic time $\sim$ λ^{-1} of the exponential divergence of two points moving along nearby trajectories;

(2) the presence of a minimal distance $d_{\min}$ between two trajectories owing to a finite spatial resolution of measurements δ, $d_{\min} \geq \delta$;

(3) the ratio of the characteristic scale R_{ch} of the velocity field variations to the resolution δ is not too large, moreover there are some regions where $R_{ch} \simeq \delta$;

(4) we cannot measure the velocity field of the overall disk but only a part of the disk.

To overcome the first difficulty we use the mentioned above property of the stationarity of the velocity field.

The second and the third difficulties restrict an allowable maximal length of the trajectory (and thus a maximal time T of the calculation of the LCN), when one uses only two neighbouring trajectories. Indeed, as the initial separation cannot be chosen arbitrarily small, the separation $d(t)$ becomes comparatively large after a short time period t so that the condition $d(t) \ll R_{ch}$ is not fulfilled, and hence the dynamics of $d(t)$ cannot be described by the liniarized equations.

To overcome these difficulties we use a method proposed by Casartelli *et al.* (1976) and described in the well-known monograph by Lichtenberg and Lieberman

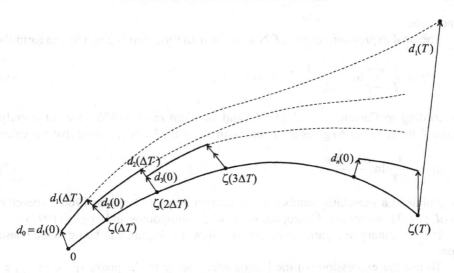

Figure 9. The method of the LCN calculation. The reference trajectory is denoted by the longest solid line and some auxiliary trajectories by dashed curves with short regions of solid curves. The minimal separation between the reference and the auxiliary trajectories is $d_0 \sim \delta$. For the sake of simplicity of notation, we omitted the symbol of the vector over d.

(1983) (see also Chirikov, Izrailev and Tayurskii, 1973; Voglis and Contopoulos, 1994; Contopoulos and Voglis, 1997).

The longest solid line in the Figure 9 denotes a 'reference' trajectory, for which we calculate the LCN. For a given reference trajectory we choose the initial separation vector $\vec{d}_1(0) = \vec{d}_0$ ($d_0 \simeq \delta$) between two close points – on the basic trajectory and on the first auxiliary trajectory (auxiliary trajectories are shown by dashed lines with small solid pieces in Figure 9). Then we calculate the evolution of $\vec{d}(t)$ under the condition $d(t) \ll R_{ch}$. At the time $t = \Delta T \equiv T/n$ we find the separation $d_1(\Delta T)$, which gives the first estimate of the LCN:

$$\lambda^{(1)} = \frac{1}{\Delta T} \ln \frac{d_1(\Delta T)}{d_0} . \tag{3}$$

After the first step, a renormalization of the vector $\vec{d}_1(\Delta T)$ is required. It gives the initial separation for the second time step:

$$\vec{d}_2(0) = \frac{\vec{d}_1(\Delta T)}{d_1(\Delta T)} d_0 . \tag{4}$$

Using the reference and the second auxiliary trajectories we obtain the second estimate of the LCN:

$$\lambda^{(2)} = \frac{1}{2\Delta T} \sum_{i=1}^{2} \ln \frac{d_i(\Delta T)}{d_0} , \tag{5}$$

and so on.

The final expression for the LCN after the n-th time step is given by the formula

$$\lambda^{(n)} = \frac{1}{T} \sum_{i=1}^{n} \ln \frac{d_i(\Delta T)}{d_0}, \qquad T \equiv n\Delta T. \tag{6}$$

According to Casartelli et al. (1976) and Benettin et al. (1976), for sufficiently large T the relation (6) gives a reliable estimate of the LCN (1). Note that the value

$$a_i = \frac{1}{\Delta T} \ln \frac{d_i(\Delta T)}{d_0}. \tag{7}$$

is similar to a stretching number in the terminology of Oseledec 1968, Froeschlé et al. (1993), Voglis and Contopoulos (1994), Contopoulos and Voglis (1997).

It is customary to regard the trajectories shown in Figure 9 as being in the phase space.

To use the expression (6) the Riemannian metric of the phase space (x, y, z, v_x, v_y, v_z) should be specified. In the mathematical literature a proof on the independence of the LCN from the Riemannian metric can be found in monographs published more than 40 years ago (for some references see, e.g., Oseledec 1968). To avoid mixing of the spatial and velocity coordinates we suggest the following metric[3]:

$$d = \sqrt{(x' - x)^2 + (y' - y)^2}. \tag{8}$$

Note that for some trajectories we recalculated the LCN using other expressions for d and obtained quite close results. For example, we used

$$d = |x' - x|; \tag{9}$$
$$d = |y' - y|; \tag{10}$$
$$d = \sqrt{(v'_x - v_x)^2 + (v'_y - v_y)^2}; \tag{11}$$
$$d = |v'_x - v_x|; \tag{12}$$
$$d = |v'_y - v_y|. \tag{13}$$

Thus we show the LCN independence from the Riemannian metric for real systems by the particular example of the galaxy NGC 3631.

Because of the third and the forth difficulties mentioned above, a trajectory cannot be traced for arbitrarily large time. First, a trajectory can eventually leave the area, for which the velocity field is defined. Second, a trajectory may come to the region where the characteristic scale of the velocity field variations is of the order of the spatial resolution of the velocity data, $R_{ch} \simeq \delta$. The further course of the trajectory cannot be calculated reliably. Such regions are 'forbidden' for the method used to find the LCN.

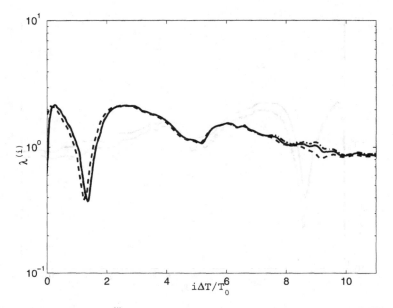

Figure 10. The behaviour of $\lambda^{(i)}$ with $i\Delta T/T_0$ for different numbers of auxiliary trajectories n calculated for a reference trajectory near the separatrix of the anticyclone. $n = 10000$ (solid line); $n = 1000$ (dash-dotted line); $n = 100$ (dotted line).

The limited length of the trajectory used in the calculations poses a question on the accuracy of the LCN determination. We consider the results to be reliable when:

$$\frac{d_1(T)}{d(0)} \equiv \exp(\lambda^{(n)}T) \gg 1, \qquad or \qquad \xi \equiv \lambda^{(n)}T \gg 1. \qquad (14)$$

Sometimes, the trajectories are so short that the condition (14) is not fulfilled. According to Voglis and Contopoulos (1994), Contopoulos and Voglis (1997) the reliability can be improved, if one takes a set of trajectories instead of one. In this case the LCN is calculated as follows:

$$\lambda^{(n)} = \frac{1}{N} \sum_{k=1}^{N} \lambda_k^{(n)}, \qquad (15)$$

where $\lambda_k^{(n)}$ is calculated according to the formula (6), using the k-th trajectory as a reference one, N is the total number of the basic trajectories. The reliability condition turns into

$$\eta \equiv \xi N \equiv \lambda^{(n)}TN \gg 1. \qquad (16)$$

Figure 10 represents the dependence of $\lambda^{(i)}$ on $i\Delta T/T_0$, $T_0 = 7.5 \cdot 10^7$ years, for various n calculated for the reference trajectory near the anticyclone (marked

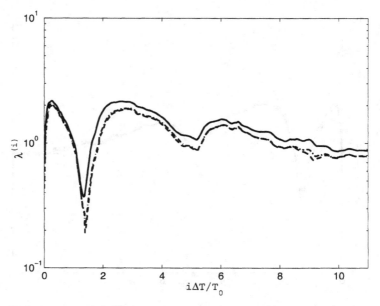

Figure 11. The behaviour of $\lambda^{(i)}$ for different initial separations d_0: 1.4 arcsec (solid) 0.14 (dash-dotted), 0.014 (dotted).

by a thick line in Figure 8). For this trajectory, $\lambda^{(n)} = 0.87\ T_0^{-1}$ and $\xi = 9.57$. As it is follows from the figure, the behaviour of $\lambda^{(i)}$ has a weak dependence on the number of the auxiliary trajectories n. The initial separation d_0 in the calculations equals to 1.4 arcsec, that is of the order of the angular resolution of the velocity data.

It would be interesting to note that for the anticyclones the LCN, calculated using different metrics, deviate very little. Namely, for the metric (8) $\lambda^{(n)} = 0.8768T_0^{-1}$, for the metrics (9) $\lambda^{(n)} = 0.8931T_0^{-1}$, for the metric (9) $\lambda^{(n)} = 0.8513T_0^{-1}$.

We found that the LCN for the given velocity field does not practically change with the reduction of the initial separation d_0. This is seen in Figure 11 where the behaviour of $\lambda^{(i)}$ for different initial separations is shown. This insensibility implies that even $d_0 \sim \delta$ is small enough to hold the condition $d(t) \ll R_{ch}$ and suitable for the LCN calculation.

For the cyclone we obtained $\lambda^{(n)} = 1.04\ T_0^{-1}$ (see Figure 12) and $\xi = 6.24$. For the metric (8) $\lambda^{(n)} = 1.0380T_0^{-1}$, for the metric (9) $\lambda^{(n)} = 1.0884T_0^{-1}$, for the metric (9) $\lambda^{(n)} = 0.8892T_0^{-1}$.

Due to the relatively short trajectories, the saddle point needs a special approach. We consider in Figure 8 the set of (ten) trajectories, with equally spaced initial points on the ordinate axis between $y = 18.5''$ and $y = 20.5''$. The behaviour of $\lambda^{(i)}$ calculated according to (15) is shown in Figure 13. The LCN value is

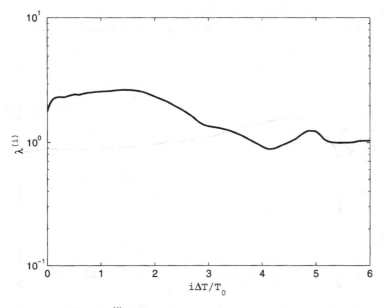

Figure 12. The behaviour of $\lambda^{(i)}$ for the reference trajectory near the separatrix of the cyclone.

$\lambda = 0.79\ T_0^{-1}$. Taking into account that in this case $T = 2\ T_0$, one can obtain that $\eta = 15.8$, i.e. the result is sufficiently reliable.

As it follows from Figures 10–13 the LCN for the trajectories in the vicinity of the anticyclones, cyclones and the saddle points are positive, i.e. these trajectories are chaotic.

In Figure 14 we can see an example of the LCN ≈ 0 for the trajectories in the vicinity of the galactic center. These trajectories are regular.

The mentioned definition of the values of the stretching numbers a_i allows to find their spectra for real objects. According to the definition (Voglis and Contopoulos, 1994; Contopoulos and Voglis, 1997), the spectrum of the stretching numbers is

$$S(a, x_0, y_0) = \lim_{N \to \infty} \frac{1}{N} \frac{dN(a)}{da}, \tag{17}$$

where $dN(a)$ is the number of appearances of the stretching number a_i in the interval $(a, a + da)$.

In Figure 15 we show the spectrum for the stochastic trajectory in the vicinity of the antycyclone, for which the LCN calcultaion is presented in Figure 10. Figure 16 shows the respective results for the regular trajectory of Figure 14 in the central part of the disk.

As follows from (6), (7) and (17) the LCN is the first moment of the spectrum of the streching numbers. As one can see the spectra for the chaotic orbit (Figure 15) and of the regular orbit (Figure 16) have this property. The spectrum for the regular

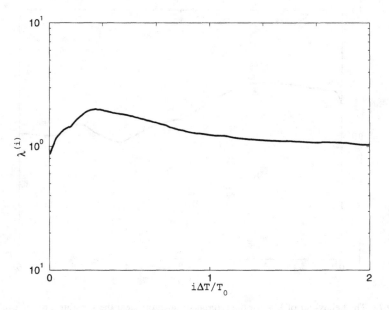

Figure 13. The behaviour of $\lambda^{(i)}$ for the set of reference trajectories near a saddle point.

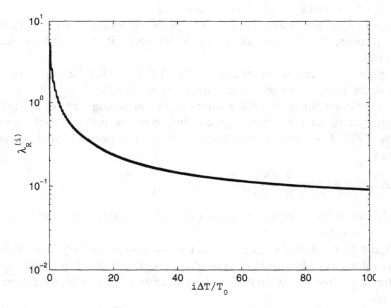

Figure 14. The behaviour of $\lambda^{(i)}$ for a regular trajectory near the galactic center, chosen as an example.

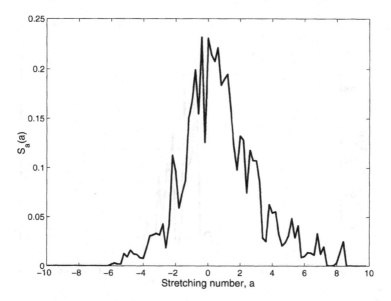

Figure 15. The spectrum for the stochastic trajectory in the vicinity of the anticyclone (see Figure 8).

orbit has two peaks. The left peak (at $a < 0$) is higher and thinner that the right one at $a > 0$. As the result the corresponding LCN is zero. Alternatively the spectrum for the chaotic trajectory (Figure 15) is clearly non-symmetrical with respect to the center of the distribution, and is shifted to positive a's.

We plan to analyse the spectra in more detail in a forthcoming paper (Fridman, Khoruzhii, and Polyachenko 2003).

To conclude, let us note that the velocity field $(\bar{v}_x, \bar{v}_y, \bar{v}_z)$ represented by arrows squares and asterisks in Figure 7 , is a result of averaging of the three-dimensional field $[v_x(\vec{r}), v_y(\vec{r}), v_z(\vec{r})]$ over the disk width (with some 'weight coefficient', much lower for the more distant half of the disk).

The functions $v_x(\vec{r})$ and $v_y(\vec{r})$ depend on z to a rather small extent. If the chosen family of streamlines of the field $(\bar{v}_x, \bar{v}_y)$ is characterized by the positive λ, i.e. the divergence is exponential, then the real (three-dimensional) streamlines are bound to diverge exponentially – else we wouldn't get the exponentially diverging trajectories as a result of averaging over z. Therefore the fact of λ being positive for the averaged over z velocity field $(\bar{v}_x, \bar{v}_y)$ means that for the corresponding family of streamlines in the real three-dimensional velocity field, where, unlike the two-dimensional case[4], the streamlines can get 'entangled', the value of λ is also positive. This sort of 'mixing' of the streamlines in the coordinate space, in appearance much similar to that which happens to the trajectories of the particles in the phase space of the Hamiltonian systems, might lead to chaos (Zaslavski *et al.* 1991).

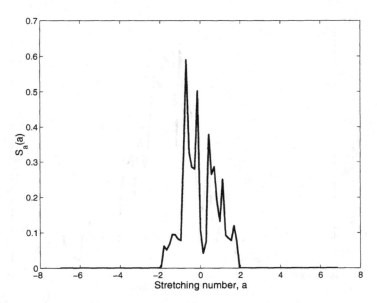

Figure 16. The spectrum for the regular in the central part of the disk (see Figure 8).

Notes

1. As shown earlier (Sakhibov and Smirnov, 1989; Canzian, 1993; Fridman *et al.*, 1997), if the circular velocity of the gas in a galaxy is perturbed by a two-armed spiral pattern, this must lead to the appearance of the first and the third Fourier harmonics ($m_{\text{obs}} = 1$ and 3) in the azimuthal distribution of the observed line-of-sight velocity. In addition, the second harmonic ($m_{\text{obs}} = 2$) may also appear if the density wave induces vertical oscillations of the gas (Fridman *et al.*, 1997, 1998).

2. As pointed out in Fridman *et al.* (1998, 2001a) the reasons for the differences between radio and optical estimates of the amplitudes of the second Fourier harmonics on the one hand, and of the first and the third harmonics on the other hand, should be different. In the latter case, the difference may be caused by the low resolution of the radio data. This is well illustrated in Figure 4 b, where the amplitudes were calculated after smoothing of the optical velocity field to a resolution of $14''$, close to that of the radio data. The amplitude squares of the first and third harmonics in Figure 4 b are one-third/one-fifth of those in the original Hα data. So, as follows from Figure 4, the histograms of the Hα smoothed data occupy an intermediate position between the histograms of the original Hα and HI data. The second harmonic of the smoothed Hα line-of-sight velocity field is naturally much higher than the second harmonic of the HI data.

3. Note that in our case the Riemannian metric is:
$$ds = \sqrt{g_{ik}dx^i dx^k} = \sqrt{g_{11}dx^2 + g_{22}dy^2 + g_{33}dz^2 + g_{44}dv_x^2 + g_{55}dv_y^2 + g_{66}dv_z^2}.$$ According to the metric definition if any $(dx^i)^2 > 0$ we should have $ds > 0$. It means that in the case of (8) the values of g_{33}, g_{44}, g_{55} and g_{66} should be considered as non zero, but positive and infinitesimal, in the case of (9) the values of g_{22}, g_{33}, g_{44}, g_{55} and g_{66} should be similar above, etc.

4. Remember that in hydrodynamics, the intersection of the streamlines is impossible.

Acknowledgements

Authors are grateful to B. Chirikov, G. Contopoulos, Y. Fridman, V. Oseledec, M. Rabinovich, R. Sagdeev, and Ja. Sinai for numerous and very fruitful discussions. This work was performed with partial financial support of RFBR grant N 99-02-18432, 'Leading Scientific Schools' grant N 00-15-96528, and 'Fundamental Space Researches. Astronomy' grants: N 1.2.3.1, N 1.7.4.3.

References

Arp, H.: 1996, Atlas of Peculiar Galaxies. *ApJS* **14**, 1.

Bennetin, G., Galgani, L., and Strelcyn, J.: 1976, 'Kolmogorov entropy and numerical experiments', *Phys. Rev. A.*, **14**, 2338.

Boeshaar, G. and Hodge, P. W.: 1977, 'H II Regions and the Spiral Structure of NGC 3631', *ApJ* **213**, 361.

Canzian, B.: 1993, 'A new way to locate corotation resonances in spiral galaxies', *ApJ* **414**, 487.

Casartelli, M., Diana, E., Galgani, L., and Scotti, A.: 1976, *Phys. Rev.* **A13**, 1921.

Chirikov, B. V., Izrailev, F. M., Tayurski, V. A.: 1973, *Comput. Physics Commun.* **5**, 11.

Contopoulos, G. and Voglis, N.: 1997, 'A Fast Method for Distinguishing Between Ordered and Chaotic Orbits', *A&A* **317**, 73.

England, M. N., Hunter, J. H. Jr., and Contopoulos, G.: 2000, 'Gaseous Vortices in Barred Spiral Galaxies', *Astrophys. J.*, **540**, 154.

Fridman, A. M. and Polyachenko, V. L.: 1984, *Physics of gravitating systems*, v.1,2, Springer-Verlag, New York, Berlin, Heidelberg, Tokyo.

Fridman, A. M., Khoruzhii, O. V., Lyakhovich, V. V., Avedisova, V. S., Sil'chenko, O. K., Zasov, A. V., Rastorguev, A. S., Afanasiev, V. L., Dodonov, S. N., and Boulesteix, J.: 1997, 'Spiral-Vortex Structure in the Gaseous Disks of Galaxies', *Astroph. Space Sci.* **252**, 115.

Fridman, A. M., Khoruzhii, O. V., Zasov, A. V., Sil'chenko, O. K., Moiseev, A. V., Burlak, A. N., Afanasiev, V. L., Dodonov, S. N., and Knapen, J. H.: 1998, 'Vertical motions in the gaseous disk of the spiral galaxy NGC 3631', *Astron. Lett.*, **24**, 764.

Fridman, A. M., Khoruzhii, O. V., Polyachenko, E. V., Zasov, A. V., Sil'chenko, O. K., Afanasiev, V. L., Dodonov, S. N., and Moiseev, A. V.: 1999, 'Giant cyclones in gaseous disks of spiral galaxies', *Phys. Lett.* **A264**, 85.

Fridman, A. M., Khoruzhii, O. V., Polyachenko, E. V., Zasov, A. V., Sil'chenko, O. K., Moiseev, A. V., Burlak, A. N., Afanasiev, V. L., Dodonov, S. N., and Knapen, J. H.: 2001a, 'Gas motions in the plane of the spiral galaxy NGC 3631', *MNRAS* **323**, 651.

Fridman, A. M., Khoruzhii, O. V., Lyakhovich, V. .V., Sil'chenko, O. K., Zasov, A. V., Afanasiev, V. L., Dodonov, S. N. and Boulesteix, J.: 2001b, 'Restoring the full velocity field in the gaseous disk of the spiral galaxy NGC 157', *A&A*, **371**, 538.

Fridman, A.M., Khoruzhii, O.V., and Polyachenko, E.V.: 2003 (to appear).

Hodge, P. W.: 1982, 'The billion-year-old clusters of the Magellanic Clouds', *ApJ*, **256**, 447.

Knapen, J. H.: 1997, 'Atomic hydrogen in the spiral galaxy NGC 3631', *MNRAS* **286**, 403.

Lichtenberg, A.J. and Lieberman, M.A.: 1983, 'Regular and stochastic motion', Springer-Verlag, New York, Heidelberg, Berlin.

Lindblad, B. and Langebartel, R. G.: 1953, 'On the dynamics of stellar systems', *Stockholms Observatoriums Annaler* B-d. 17, **6**, 61.

Lynden-Bell, D.: 1996, 'Formation and Evolution Mechanisms of Barred Spiral Galaxies', in *Barred galaxies and circumnuclear activity*: proceedings of the Nobel Symposium 98, Sandquist, A. and Lindblad, P. O. (eds.), Springer-Verlag, Berlin, Heidelberg, New-York, pp. 7–17.

Oseledec, B. I.: 1968, 'Multiplicative Ergotic Theorem. Lyapunov Characteristic Numbers for Dynamical Systems', *Tr. Mosk. Mat. Obsch.*, **19**, 179, (Trans. *Mosc. Math. Soc.*, **19**, 197, 1968).

Rozas, M., Beckman, J. E., and Knapen, J. H.: 1996, 'Statistics and properties of H II regions in a sample of grand design galaxies I. Luminosity functions', *A&A*, **307**, 735.

Sakhibov, F. Kh. and Smirnov, M. A.: 1989, 'Noncircular gas motions in the spiral galaxies NGC 3031, NGC 2903, and NGC 925', *Sov. Astron.*, **33**, 476.

de Vaucouleurs, G., de Vaucouleurs, A., Corwin, H. G. Jr., Buta, R. J., Paturel, G., and Fouque, P. *Third Reference Catalogue of Bright Galaxies*. Springer-Verlag, Berlin, Heidelberg, New York.

Voglis, N. and Contopoulos, G.: 1994, 'Invariant Spectra of Orbits in Dynamical Systems', *J. Phys.*, **A27**, 4899.

Zaslavsky, G. M., Sagdeev, R. Z., Usikov, D. A., and Chernikov, A. A.: 1991, 'Weak chaos and quasi-regular patterns'. *Cambrige Univ. Press*.

OBSERVING CHAOS IN EXTERNAL SPIRAL GALAXIES

P. GROSBØL

European Southern Observatory, Karl-Schwarzschild-Str. 2, D-85748 Garching, Germany
(E-mail: pgrosbol@eso.org)

Abstract. The feasibility of observing chaotic behavior in the stellar component of spiral galaxies is discussed. Three sources for development of chaos are considered namely: steep potential gradients, resonances and growing spiral perturbations. Several regions where chaos could be expected are identified such as the very central region, the end of the bar, the start of the main spiral pattern and the termination of strong spiral arms.

The main observational signature is likely to be an increased velocity dispersion while multiple peaks in the velocity profile due to bifurcation of the main family of periodic orbits near resonances could be viewed as an indicator of increased stochasticity. It is non-trivial to distinguish between a higher velocity dispersion due to chaotic motions and non-periodic orbits trapped around the central family of stable periodic orbits. This requires a good dynamic model which can be obtained by combining near-infrared K surface photometry maps and kinematic information.

The ESO VLT 8 m unit telescopes were taken as a reference to judge if it is feasible to observe chaos in disk galaxies with current state-of-the-art equipment. Whereas surface photometry map easily can be obtained with smaller telescopes, detailed line-of-sight velocity profiles from absorption lines are difficult to observed below an isophotal level of $I \approx 20$ mag/$\square''$ even with an 8 m class telescope. This suggests that it would be possible to observe chaotic behavior in spiral galaxies out to the end of the bar or start of the main spiral pattern but not further out in the spiral arms.

1. Introduction

It is relative simple to create chaotic behavior in numerical model of dynamic systems. This can be achieved by increasing perturbations of the potential over a certain limit (Contopoulos, 1983a). The existence of chaos in analytic and numeric models does not necessarily prove that chaos also is an important feature in real physical systems as models do not represent the real world in full detail. Current numerical N-body simulations are also too rough to accurately follow the evolution of galactic system. Thus, it is important to get more direct observational evidence for the importance of chaos in dynamic systems like galaxies.

The present paper considers where one would expect to find chaos in galaxies and what the possible observational indicators would be. The main emphasis is placed on disk galaxies because projection effects are less important compared to more spherical systems like elliptical galaxies. Finally, some case studies are made to verify if it is feasible to make such observations with todays state-of-the-art

*Based on observations collected at the European Southern Observatory, La Silla, Chile; Programs: 63.N-0343 and 66.N-0257

facilities like the VLT or one would need to wait for the next generation of even larger telescopes.

2. Observing Chaos

It is intrinsically difficult to observe chaotic behavior of stellar orbits in galactic systems since, in general, the dynamic time scale is long and only integrated quantities can be measured (e.g. surface density and line-of-sight velocity). Further, different dynamical processes may, in the course of the evolution of the system, give raise to the same measure of an integrated property e.g. a higher velocity dispersion could be caused by general heating in the disk or a transition to chaos.

Another problem is that chaotic regions often will have a more featureless phase space than regions with ordered motions although a lack of significant features (e.g. strong azimuthal perturbations) also could be intrinsic to the system (e.g. a hot disk which is stable against perturbations). In a scenario where chaos is introduced by unstable growing perturbations, it is not expected that a sudden change occur but rather that the fraction of chaos slowly increases as the perturbation becomes non-linear. This suggests that one will find a mixture of ordered and chaotic behavior which requires a careful analysis based on a dynamic model of the system to be disentangled. In the section below, the locations where one would expect chaos in spiral galaxies are discussed with indications of which features could be observed.

2.1. WHERE TO LOOK?

Excluding catastrophic events such as mergers or encounters of galaxies, main features which may cause a transition from ordered to chaotic stellar motions are high gradient in the underlying potential, existence of resonances and strong perturbations. The effects associated to these features can be summarized as follows:
High Gradients in potential: Many galaxies have a very sharply raising rotation curve in their central region which suggests a strong central mass concentration (e.g. a black hole). Stars reaching the center on highly eccentric, radial orbits will tend to be scattered due to the high gradient of the central potential as seen in N-body experiments (Sellwood and Moore, 1999). Since disk stars on planer orbits may be scattered out of the plan, it will be very difficult to distinguish chaotic orbits generated in this way from those which already exist in the bulge.
Main resonances: Stars moving on near circular orbit in resonance regions where a perturbation such as a spiral density wave (Lin and Shu, 1964) effects them periodically are more likely to exhibit a chaotic behavior that stars in other regions (Contopoulos and Grosbøl, 1989). For a two armed spiral wave, the most important resonances are Inner Lindblad Resonance (ILR), 4:1 resonance and Co-Rotation (CR) while the Outer Lindblad Resonance (OLR) often will be located at too large radii to be easy to observe. Resonance crowding near CR will also generate stochastic behavior even at low perturbations (Contopoulos, 1983b).

In barred galaxies, one would expect to see an increased level of chaos in the region around CR which is normally expected just beyond the end of the bar (Contopoulos, 1980). For normal spirals, stochasticity is expected just inside the main spiral pattern, where ILR is assumed, and in the case of strong spirals close to their termination corresponding to either 4:1 or CR (Contopoulos and Grosbøl, 1986). For galaxies in which bar and spiral have different pattern speed and are coupled through resonances (Sellwood and Sparke, 1988), stochastic behavior would be expected in the interface region between the two components.

Strong perturbations: Perturbations with growing modes will eventually reach an amplitude which makes non-linear effects important. As a perturbation grows stronger stable orbits may become unstable (typically starting in the resonance regions) and thereby increase the amount of stochasticity in the system. The degree of chaos created in this way will depend on the efficiency with which non-linear effect can stop the growth of the perturbation.

In addition to the existence of such features in a galaxy, the dynamical time scale associated to them is important. In the inner regions of galaxies with a short time scale it is more likely to observe a transition from ordered to chaotic motion than at large radii where chaotic behavior rather may be a result of the region never reaching a stable configuration.

2.2. WHAT TO EXPECT?

A chaotic region will in general exhibit fewer features in phase space than one occupied by ordered motions. This suggests that regions dominated by chaotic motions frequently will have a smoother mass distribution and a larger velocity dispersion than ones with more order. Since the general lack of features could be intrinsic to the galaxy, a signature of chaos would be a small, well defined, radial region with smaller azimuthal perturbations and higher velocity dispersion than the neighboring ones. However, in a real galaxy the intrinsic velocity dispersion of stars courses most of them to travel on non-periodic orbits trapped around the main family of stable periodic orbits in the disk. It may therefore be difficult to separate the integrated effect of chaotic motions from those of non-periodic orbits.

It is reasonable simple to identify chaos in models (see e.g. Voglis *et al.*, 1999) where one can study the behavior of individual orbits over time. The application of such results to observations of galaxies is non-trivial due to the wide range of stars on different orbits which contribute to the light observed in a given resolution element. The integration along the line of sight will also add other components due to the distribution of stars perpendicular to the disk and galactic rotation. More detailed studies of the dynamics are critical to fully understand and diagnose observable effects of an increased fraction of chaos in galactic systems.

In regions with a significant fraction of chaotic motion, one would expect to find a stellar component with a higher velocity dispersion than that expected for the normal disk. Depending very much of the exact shape of the chaotic region in phase space one could get an asymmetric velocity profile. A careful interpretation of such asymmetries is required as they easily may be caused by other effects such as attenuation of dust in the plane of the galaxy or improper correction for the general rotation.

Since the main family of periodic orbits splits in two branches at resonances like ILR and 4:1 for strong perturbation, one may be able to observe a double peaked velocity profile, under favorable geometric conditions, as indication of a strongly perturbed region which could develop chaos.

2.3. HOW TO OBSERVE?

It is necessary to establish a fairly good dynamic model of the galaxy being investigated in order to distinguish between ordered and chaotic motions. This requires an estimate of the general potential of the system which can be obtained through its mass distribution and velocity map (Kent, 1986). The surface density map can best be acquired from images in the near-infrared K band where attenuation of dust is minimal. Using long slit spectra (LSS) to estimate the mass-to-light ratio, one can convert K surface brightness to density under the assumption of a certain dark matter fraction. Correction for population effects must still be applied since 20–30% of the light in K may originate from young objects (Rix and Rieke, 1993). A full velocity map obtained though a spectrograph with an Integral Field Unit (IFU) can also be used to obtain the potential. It is very useful to compare these two estimates to ensure that the system is quasi-stationary and has no strong streaming motions.

After a dynamic model has been made and possible locations of chaotic regions identified, detailed line-of-sight velocity profiles of these places are obtained. The profiles are then analyzed to estimate if any components with a velocity dispersion significant larger than expected for the normal disk population exist. It is important that the spectral lines used to measure the velocity profiles ensure that a well defined old stellar population is used. Control observations of regions assumed to mainly contain ordered motion must be done to ensure that the velocity profile of the normal disk is well defined.

2.4. REGIONS WITH CHAOS?

As discussed above, one indicator for chaos is a smooth appearance in phase space. In a galactic disk where most stars move on nearly circular orbits, an increasing fraction of chaotic motions would tend to reduce azimuthal perturbations. Radial regions with very small azimuthal variation in a galaxy which otherwise exhibits significant non-axisymmetric perturbations could be considered as candidates for chaotic behavior. A detailed study of the velocity distribution would be required to determine the actual fraction of chaotic to ordered motions. Two spiral galaxies, NGC 1566 and NGC 4939, observed in the K' band with SOFI on the NTT at La Silla are used as examples.

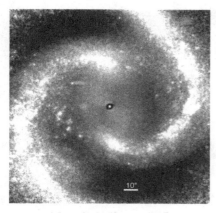
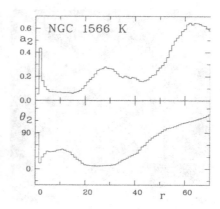

a. Map of relative perturbation   b. Radial amplitude and phase

Figure 1. Inner parts of NGC 1566 in K′ band: (a) face-on map of intensity relative to the axisymmetric disk with the bulge subtracted, and (b) radial variation of the amplitude, a_2, and phase, θ_2, of the azimuthal m = 2 Fourier component. The relative intensities inside the central 5″ cannot be trusted due to the very high gradients. Brighter areas represent intensities higher that the azimuthal average.

The grand design, spiral galaxy NGC 1566 is classified as SXS4 by de Vaucouleurs *et al.* (1991). Its inner parts are shown in Figure 1a where the K band intensity relative to the axisymmetric average is given. A bulge component was fitted and subtracted before the galaxy was de-projected. The radial variation of amplitude and phase of the bisymmetric perturbation in the disks (i.e., the azimuthal m = 2 Fourier component) are displayed in Figure 1b. A smooth oval distortion with an amplitude of 7% is seen inside a radius of 15″. There is a strong, point like source in the center which could suggest a steep, central mass gradient. Both this gradient and the smooth appearance of the bar may indicate that some chaotic motions exist. Also the region between the end of the bar and the inner start of main spiral arms around 20″ has a small azimuthal variation and may be a candidate for chaotic behavior. The very inner parts of the main spiral have a high amplitude which reaches 28% around r ≈ 28″. The patchy appearance of this feature suggests that a significant part of the K band flux originates from a young population. Shocks and chaotic motions in the gas may enhance the star formation in this region.

Another example is NGC 4939 which is classified as a SAS4 spiral (see Figure 2). The relative intensity map in Figure 2a shows a bar structure inside 18″ with many symmetric features which suggest a high fraction of ordered motion. A short set of spiral arms can be seen just outside the bar. They reaches 26″ while the main spiral pattern first starts around 33″. The region between the two spirals has a low amplitude (<4%) which may be caused by a larger fraction of chaotic orbits. It is noted that this galaxy also shows an increased amplitude in the inner part of its

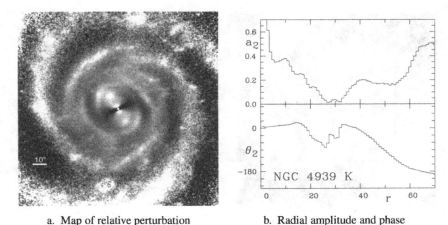

a. Map of relative perturbation   b. Radial amplitude and phase

Figure 2. Inner parts of NGC 4939 in K′ band as in Figure 1

main spiral pattern around 40″. The symmetry of the radial amplitude variation in the two spiral arms could indicate an interaction of different wave packages in the disk.

3. Observational Considerations

The need for making a dynamic model of the galaxy and obtaining accurate line-of-sight velocity profiles requires access to high quality surface photometry and spectroscopy. The requirements for such observations are discussed in the paragraphs below.

3.1. SURFACE BRIGHTNESS

An estimate of the stellar surface density is required to access the fraction of matter located in the disk relative to the total surface density derived from the rotation curve. Surface photometry in the K band gives the most reliable results (Rix and Rieke, 1993) although population effects are present. The typical surface brightness in K at the end of the bar or start of the main spiral pattern is ≈18 mag/□″ while the spiral arms in the middle of the disk may be a magnitude fainter (see Grosbøl and Patsis, 1998). Color indices in the disk are (I-K) ≈ 2 mag and (V-K) ≈ 3 mag. Accurate photometry at these levels can easily be done with 4m class telescopes where the main limitation is faint background galaxies which make it difficult to reach fainter than K ≈ 21 mag/□″.

3.2. VELOCITY PROFILE

The velocity dispersion in the stellar disk depends on the age of the stars and is in the range 40-80 km/s for the old disk population (Bottema, 1993). In order to identify an additional component with a higher dispersion originating from chaotic motions, spectra with a resolution of at least 2000 and a signal-to-noise ratio (S/N) in the range of 20–50 are required. The actual S/N needed to identify several velocity components depends on the quality of the model predictions and the number of free parameters to be determined. The velocity profiles will normally be asymmetric which makes it important to use generic decomposition methods for the analysis e.g. by Kuijken and Merrifield (1993).

Problems in correcting for systematic effect like velocity components due to the galactic rotation and motions perpendicular to the plane make it essential to consider the projection and geometry of the galaxies. In order to minimize effects due to rotation, one would tend to observe close to the minor axis of the system. Observing in visual bands, the effect of attenuation by dust also have to be considered.

4. VLT Case Study

The ESO Very Large Telescope was used as an example of a state-of-the-art ground-based observing facility to access the technically feasibility of observing chaos in spiral galaxies with current equipment. With its four 8m telescopes located at an excellent site in the Atacama desert, Chile, and a large set of instruments, it provides a good reference for what is possible at a modern observatory.

4.1. VLT INSTRUMENTS

The VLT has a very comprehensive set of instrument which covers ultraviolet to infrared wavelengths with a wide range of formats and resolutions. The instruments of most interest for the observation of chaos in external galaxies are shortly described below.

FORS is a focal reducer for visual bands with both imaging and spectroscopic modes. It can do long slit spectroscopy with a resolution of 1700.

ISAAC is an infrared imager and spectrometer with long slit option. In the near-infrared range, it has a spectral resolution of up to around 3000 which is just enough to resolve the sky OH emission lines. This allows the usage of low background, interline spectral regions for spectroscopy of faint sources.

VIMOS is a highly multiplexed, visual fiber spectrograph with imaging mode. It has an IFU with a field of 1′ and a spectral resolution of up to 2200.

FLAMES is a visual multi-object fiber spectometer with several kinds of IFU's e.g. one 7″ × 11″ field or 15 fields of 2.5″ × 3.5″. Its lower spectral resolution is around 9000.

TABLE I

Performance of VLT instruments in spectroscopy mode for a 1 hour exposure with
0.7″ seeing. The wavelength (Δλ) and velocity (Δv) resolution together with the
Point Spread Function (PSF) are given for each configuration. Finally, the number
of electron from the source and the corresponding S/N are listed as calculated
by the ESO Exposure Time Calculator for an extended source with a surface
brightness of I = 20 mag/□″.

Instrument	Mode		Δλ (nm/pix)	Δv (km/s/pix)	PSF (pix)	Source (e⁻)	S/N
FORS	LSS	600I	0.106	39	5	760	20
	LSS	1400V	0.050	29	5	515	10
	LSS	1028Z	0.068	23	5	805	18
ISAAC	LSS	MR	0.121	16	7	550	4:
VIMOS	IFU	R2150	0.061	29	5	466	17
	IFU	R1000	0.273	117	5	3478	48
FLAMES	IFU	R9000	0.020	10	3	98	7

4.2. TYPICAL INSTRUMENT PERFORMANCE

An Exposure Time Calculator (ETC) is provided by ESO (Ballester *et al.*, 1999)
for each VLT instrument to help users optimize their usage of the facility. For all
standard instrument modes, it calculates the efficiency of a given configuration and
estimates the S/N for a source as specified. In case of instruments which have not
yet been commissioned (e.g. VIMOS and FLAMES), the efficiencies are derived
from engineering estimates for the individual components.

As source, the average energy distribution from a late-type spiral galaxy with a
surface brightness of I = 20 mag/□″ or K = 18 mag/□″ was used. This corresponds
to a typical brightness at the end of the bar or the start of the main spiral pattern
which are possible places to expect chaotic behavior. An exposure time of 1 hour
and a seeing of 0.7″ were used for all estimates to make a comparison easy. The
results are given in Table I for several configuration where the number of detected
electrons and S/N are given for one pixel in the dispersion direction while integrated over the fiber or seeing disk perpendicular to it. In the case of ISAAC, the
S/N estimate is rather uncertain due to the strong OH emission lines in the K band
spectrum. If such OH lines interfere with the absorption lines used for measuring
the velocity profile, the S/N may be reduced to close to unity.

It is possible to increase the S/N of the resulting velocity profile further by
using several spectral lines (e.g. Mg b triplet at 517 nm). The instruments with

IFU's provide more spatial channels which also can be averaged to improve the final result.

4.3. LIMITS OF 8 M TELESCOPES

From the estimates given in Table I, it is clearly possible to obtain spectra at the end of the bar with sufficient spectral resolution and a S/N in the range of 20–50 with the VLT although it is marginal with some instruments (e.g. FLAMES and ISAAC). This suggests that regions inside an isophotal level of $I = 20$ mag/$\square''$ could be observed with sufficient accuracy to identify strong chaotic behavior. The central region, the bar and the start of the main spiral pattern in a typical spiral galaxy could be investigated while adequate velocity information for the spiral arms themselves would be very difficult to obtain.

5. Conclusions

The prime candidate locations for finding chaotic behavior in spiral galaxies are the central region, the end of the bar and major resonances between stellar epicyclic motion and a spiral density wave such as ILR, 4/1 and CR. A main indicator for chaos is the velocity dispersion which is expected to be larger than for regions with ordered motions. Further, a reduction of the azimuthal amplitude variation is likely. As such features also could be intrinsic to the galaxy care must be taken in the interpretation. More studies of the dynamics of galactic systems are required to clearly understand the observational implications of chaos. A significant issue is to separate the contributions to the velocity profile from chaotic motions and non-periodic orbits trapped around stable periodic orbit. The distinction between these different types of contribution requires a good dynamic model of the galaxy which can be obtained through K band maps and general kinematic information (e.g. HI velocity width or rotation curve). Great care must be exercised in the analysis to properly correct for other effects such as attenuation by dust and galactic rotation.

Spectra with sufficient spectral resolution (>2000) and S/N (>20) to measure velocity profiles from absorption line can be obtained with 8 m class telescopes down to an isophotal level of at least $I = 20$ mag/$\square''$. This corresponds to a typical surface brightness at the end of the bar or start of the main spiral pattern. Thus, it should be possible to detect chaos in spiral galaxies out to this distance with current state-of-the-art equipment like the VLT while the study of chaos in the spiral structure itself would require even more powerful instruments.

References

Ballester, P., Disarò, A., Dorigo, D., Modigliani, A., Pizarro de la Iglesia, J. A: 1999, *ESO Messenger* **96**, 19.

Bottema, R.: 1993, *Astron. Astrophys.* **275**, 16.
Contopoulos, G.: 1980, *Astron. Astrophys.* **81**, 198.
Contopoulos, G.: 1983, *Astron. Astrophys.* **117**, 89.
Contopoulos, G.: 1983, *Astrophys. J.* **275**, 511.
Contopoulos, G. and Grosbøl, P.: 1986, *Astron. Astrophys.* **155**, 11.
Contopoulos, G. and Grosbøl, P.: 1989, *Astron. Astrophys. Rev.* **1**, 261.
de Vaucouleurs, G., de Vaucouleurs, A. and Cowien, Jr., H. G.: 1991 *Third reference catalog of bright galaxies*, Springer, New York.
Grosbøl, P. J. and Patsis, P. A.: 1998, *Astron. Astrophys.* **336**, 840.
Kent, S. M.: 1986, *Astron. J.* **91**, 1301.
Kuijken, K. and Merrifield, M. R.: 1993, *Mon. Not. R. Astron. Soc.* **264**, 712.
Lin, C. C. and Shu, F. H.: 1964, *Astrophys. J.* **140**, 646.
Rix, H.-W. and Rieke, M. J.: 1993, *Astrophys. J.* **418**, 123.
Sellwood, J. A. and Moore, E. M.: 1999, *Astrophys. J.* **510**, 125.
Sellwood, J. A. and Sparke, L. S.: 1988, *Mon. Not. R. Astron. Soc.* **231**, 25.
Voglis, N., Contopoulos, G. and Efthymiopoulos, C.: 1999, *Cel. Mech. Dyn. Astron.* **73**, 211.

SPECTRAL ANALYSIS OF ORBITS VIA DISCRETE FOURIER TRANSFORMS

C. HUNTER

Department of Mathematics, Florida State University, Tallahassee, Florida 32306-4510, USA
(E-mail: hunter@math.fsu.edu)

Abstract. We give some simple and direct algorithms for deriving the Fourier series which describe the quasi-periodic motion of regular orbits from numerical integrations of those orbits. The algorithms rely entirely on discrete Fourier transforms. We calibrate the algorithms by applying them to some orbits which were studied earlier using the NAFF method. The new algorithms reproduce the test orbits accurately, satisfy constraints which are consequences of Hamiltonian theory, and are faster. We discuss the rate at which the Fourier series converge, and practical limits on the degree of accuracy that can reasonably be achieved.

1. Introduction

This work develops efficient and accurate methods for computing Fourier expansions of orbits in potentials of galactic type. Their ultimate purpose is for use in constructing stellar dynamic models. Efficient methods are needed because galactic models require the use of large numbers of orbits. The methods described here are designed specifically for regular orbits, and may be useful also with nearly regular orbits whose characteristics change slowly with time. This is not to suggest that regular and near-regular orbits are the only ones of interest in galactic dynamics – they are not. But regular orbits can be an important component of galaxies, and this work is relevant to them at least.

The methods to be described use Fourier analysis. Fourier analysis of orbits in galactic-type potentials was pioneered by Binney and Spergel (1982, 1984). There has been a recent upsurge of interest as a result of the work of J. Laskar (1990, 1993, 1999) who has developed accurate numerical methods of Fourier analysis which are known by the acronym NAFF (Numerical Analysis of Fundamental Frequencies). These methods were developed originally for solar system phenomena, including mildly chaotic ones, but they were also applied by Papaphilippou and Laskar (1996, 1998) to potentials of galactic type. Other applications of NAFF to cuspy triaxial galactic-type potentials have followed, including those by Carpintero and Aguilar (1998), Valluri and Merritt (1998), Wachlin and Ferraz-Mello (1998), and Merritt and Valluri (1999). Copin, Zhao and de Zeeuw (2000) have that smooth orbital densities can be derived from Fourier representations of orbits obtained using NAFF, and have illustrated the method on a Stäckel potential.

Space Science Reviews **102**: 83–99, 2002.
© 2002 *Kluwer Academic Publishers. Printed in the Netherlands.*

NAFF begins with the numerical integration of an orbit, and the recording of its phase space coordinates at a sequence of equally spaced time steps. A discrete Fourier transform (DFT) is applied to this data, and the most prominent frequency is identified as that belonging to the largest Fourier coefficient [c.f. Binney and Spergel (1982)]. This provides an initial estimate of the most prominent frequency. This estimate is refined by the use of Fourier integrals and window functions. The refined frequency is that which maximizes the Fourier integral which represents the amplitude associated with that frequency. This component is then subtracted out, and subsequent Fourier components are identified sequentially in order of diminishing prominence, and subtracted in turn.

This work provides an alternative way of performing the Fourier analysis entirely with discrete transforms. It is natural to seek such a method because the continuous orbit is discretized by the initial numerical integration, and it is hard to see what can be gained by switching to Fourier integrals subsequently. We describe our method in Section 3, after first reviewing some essential elements of Hamiltonian Dynamics in Section 2. We calibrate it in Section 4 by applying it to three cases discussed in detail by Papaphilippou and Laskar (1996). In Section 5 we compare our method with that of Laskar, and give our conclusions. An appendix justifies a key aspect of our method of determining frequencies of Fourier components in a time series.

2. Hamiltonian Dynamics

A regular orbit in a system with n degrees of freedom lies on an n-torus in phase space (Binney and Tremaine 1987). It is generally quasi-periodic, and has n fundamental frequencies which we label as v_1 to v_n. The position vector of the orbit has the Fourier representation

$$\mathbf{x}(t) = \sum_{\mathbf{k}} \mathbf{X}_{\mathbf{k}}(\mathbf{I}) e^{i(\mathbf{k} \cdot v)t}. \tag{1}$$

where v is the n-vector of the fundamental frequencies, and $\mathbf{k}$ is an n-vector with integer components. Summation is over all such integer vectors $\mathbf{k}$. The Fourier coefficients $\mathbf{X}_{\mathbf{k}}$ are functions of the actions $\mathbf{I}$. Actions and their conjugate angles ϕ form a canonical set of variables, in terms of which the Hamiltonian is $H(\mathbf{I})$ and independent of the angles. The value of the action vector is constant for each specific orbit, Consequently the angle ϕ_j increases uniformly with time at the rate $v_j = \partial H / \partial I_j$. The phase space variables of position and velocity have Fourier representations in the action-angle variables $(\mathbf{I}, \phi)$ which are given by the equations

$$\mathbf{x} = \sum_{\mathbf{k}} \mathbf{X}_{\mathbf{k}}(\mathbf{I}) e^{i\mathbf{k} \cdot \phi}, \quad \mathbf{v} = \sum_{\mathbf{k}} i(\mathbf{k} \cdot v) \mathbf{X}_{\mathbf{k}}(\mathbf{I}) e^{i\mathbf{k} \cdot \phi}. \tag{2}$$

Here we either require that each angle ϕ_j be zero at time $t = 0$, or else absorb extra constant terms into the definitions of the Fourier coefficients $\mathbf{X}_{\mathbf{k}}(\mathbf{I})$.

Fourier series are strongly constrained by orbital symmetries, and the Hamiltonian. Specifically

$$I_j = \frac{1}{2\pi} \oint_{\mathcal{C}_j} \mathbf{v} \cdot d\mathbf{x}, \tag{3}$$

where $\mathcal{C}_j$ is a circuit around the torus of constant $\mathbf{I}$, and which is such that the angle ϕ_j increases by 2π, but other angles are unchanged. Then

$$I_j = \sum_{\mathbf{k}} k_j (\mathbf{k} \cdot v) \mathbf{X_k} \cdot \mathbf{X_{-k}} = \sum_{\mathbf{k}} k_j (\mathbf{k} \cdot v) |\mathbf{X_k}|^2, \tag{4}$$

while

$$\sum_{\mathbf{k}} k_j (\mathbf{k} \cdot v) \mathbf{X_k} \cdot \mathbf{X_{m-k}} = 0, \tag{5}$$

where $\mathbf{m}$ is a non-zero integer vector with zero j'th component (Binney and Spergel 1984).

3. Implementing NAFF the Discrete Transform Way

3.1. DISCRETE TRANSFORMS AND WINDOWS

We use a centered time interval and integrate from time $t = -T$ to $t = T$, with positions and velocities recorded at $2N$ equally spaced times in steps of $\Delta t = T/N$. The discrete Fourier coefficients of a function $f(t)$ are calculated from its tabulated values as

$$F_k = \frac{1}{2N} \sum_{n=1-N}^{N} f(n\Delta t) e^{-\pi i n k/N}, \tag{6}$$

for $k = 1 - N$ to N, and can be computed using a fast Fourier transform. We use the average $\frac{1}{2}[f(T) + f(-T)]$ when evaluating $f(T)$ in this formula (Briggs and Henson 1995). Then consistently $f(t)$ is periodic of period $2T$ in t and F_k is periodic of period $2N$ in k. We use the same window functions

$$\chi_p \left(\frac{t}{T} \right) = \frac{2^p (p!)^2}{(2p)!} \left(1 + \cos \frac{\pi t}{T} \right)^p, \tag{7}$$

as Laskar. These window functions taper to zero at the ends of the interval $[-T, T]$ with increasing rapidity for increasing values of p. They are ideally suited to the DFT. This is because

$$1 + \cos \frac{\pi t}{T} = \frac{1}{2} e^{i\pi t/N\Delta t} + 1 + \frac{1}{2} e^{-i\pi t/N\Delta t}. \tag{8}$$

Hence the DFT of $f(t)\chi_1(t/T)$ is simply derived from that of $f(T)$ as

$$F_k^{(1)} = \tfrac{1}{2}F_{k-1} + F_k + \tfrac{1}{2}F_{k+1}. \tag{9}$$

The Fourier coefficients of f multipled by window functions of any order, $f(t)\chi_p$ (t/T), can be computed iteratively with minimal effort using the recursive relation

$$F_k^{(p)} = \frac{p}{2p-1}\left[\tfrac{1}{2}F_{k-1}^{(p-1)} + F_k^{(p-1)} + \tfrac{1}{2}F_{k+1}^{(p-1)}\right], \qquad F_k^{(0)} = F_k. \tag{10}$$

3.2. ESTIMATING FREQUENCIES

The essential purpose of the Fourier analysis is to identify the elementary frequency components of $f(t)$. We now consider such a component $g(t) = e^{i\nu t}$, and its windowed Fourier coefficients. We denote the latter as $G_k(\nu)$. They are calculated from $g(t)$ in the same way as in equation (6), and we obtain

$$G_k(\nu) = \frac{\sin\theta}{2N\tan\left(\dfrac{\theta}{2N}\right)} = S(\theta), \quad \text{where} \quad \theta = \nu T - k\pi. \tag{11}$$

This equation defines the function $S(\theta)$ which is the discrete analog of the sinc function which arises with continuous Fourier transforms, and to which it indeed tends as $N \to \infty$. It is even and periodic of period $2N\pi$ in θ, and vanishes at all integer multiples of π, except for those which are also integer multiples of $2N\pi$. The windowed Fourier coefficients of $g(t)$ can also be calculated iteratively and are

$$G_k^{(p)}(\nu) = S^{(p)}(\theta) = \frac{p}{2p-1}\left[\tfrac{1}{2}S^{(p-1)}(\theta + \pi) + S^{(p-1)}(\theta)\right.$$
$$\left. + \tfrac{1}{2}S^{(p-1)}(\theta - \pi)\right], \tag{12}$$
$$S^{(0)}(\theta) = S(\theta).$$

An explicit expression for the DFT of $g(t)\chi_1(t/T)$ is

$$G_k^{(1)}(\nu) = \frac{\sin\theta}{2N\tan(\frac{\theta}{2N})\left[1 - \sin^2\left(\dfrac{\theta}{2N}\right)\Big/\sin^2\left(\dfrac{\pi}{2N}\right)\right]}, \tag{13}$$

The DFT of $g(t)$ and of $g(t)$ multiplied by the first two of the window functions (7) are shown in Figure 1. With increasing p the central peaks become successively wider while the side lobes decay increasingly rapidly so that frequencies increasingly stand out from the background. Figure 1 is plotted for the impractically small value of $N = 10$ to emphasize the periodicity of period $2N$ of all discrete Fourier

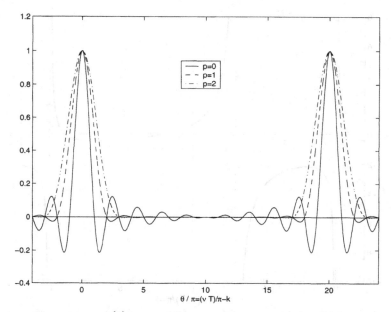

Figure 1. Fourier transforms $G_k^{(p)}(\nu) = S^{(p)}(\theta)$ of a simple exponential $e^{i\nu t}$ with no windowing ($p = 0$) and with the first two window functions (7) for $N = 10$

coefficients. As N becomes large, the discrete Fourier coefficients approach the limits

$$\lim_{N \to \infty} G_k^{(p)}(\nu) = \frac{\sin \theta}{\theta \prod_{j=1}^{p} \left(1 - \frac{\theta^2}{j^2 \pi^2}\right)}. \tag{14}$$

Although the DFT in Figure 1 are plotted for continuous ranges of their arguments, they are of course known only at discrete values of θ which are π apart in any application.

We estimate unknown frequencies ν using ratios of adjacent Fourier coefficients. This idea is due to Lanczos (1956) who used it with the $p = 1$ (Hanning) window function. It was rediscovered by Carpintero and Aguilar (1998) who used it without windowing ($p = 0$). We use it without restriction on p as follows. We look for a prominent frequency by locating a range of k values where the Fourier coefficients F_k are relatively large, and humps similar to those shown in Figure 1 develop. As Figure 2 implies, the ratios $F_{k-1}^{(p)}/F_k^{(p)}$ are positive for a limited range of k values. We select some k for which the ratio is positive and equal to C say. We

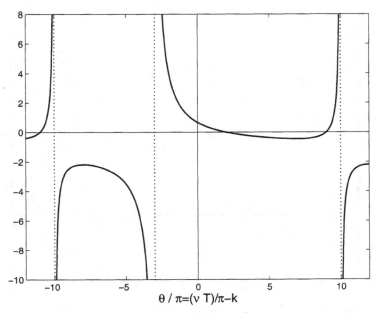

Figure 2. Ratio of two adjacent finite transforms $G^{(2)}_{k-1}(\nu)/G^{(2)}_k(\nu)$ as functions of $\theta = \nu T - k\pi$ for $N = 10$. The ratio is approximately $(p\pi - \theta)/[\theta + (p+1)\pi]$ for large N.

then find the frequency ν for which the ratio $G^{(p)}_{k-1}(\nu)/G^{(p)}_k(\nu)$ has the same ratio C. This gives us an equation

$$\frac{G^{(p)}_{k-1}(\nu)}{G^{(p)}_k(\nu)} = \frac{S^{(p)}(\theta + \pi)}{S^{(p)}(\theta)} = C, \qquad (15)$$

for θ. We show in the Appendix that it is in fact a polynomial equation of degree $2p$ in $\tau = \tan(\theta/2N)$. One need not compute all its roots, all but two of which are complex for positive C (c.f. Figure 2), because $S^{(p)}(\theta + \pi)/S^{(p)}(\theta)$ becomes negative outside the range $-(p+1)\pi \le \theta \le p\pi$ where the numerator and denominator are out of phase. Hence we need only the small real root for τ which lies in the narrow range $-(p+1)\pi/2N \le \theta/2N \le p\pi/2N$. From τ, we have θ and then the frequency $\nu = (\theta + k\pi)/T$. These estimates should become increasingly accurate with increasing p, as the sidelobes decay and individual frequencies stand out. One can easily test whether this is the case by checking for consistency between estimates at neighboring values of k, and by calculating a sequence of estimates for successive increments of p, and checking whether they converge. Table I shows the results of two analyses of the x-axis orbit discussed in Section 4.1, using the same integration interval but different spacings. The right hand set show consistent convergence to the true frequency, while the left hand set lack consistency, although getting quite close at $p = 4$,

TABLE I

Two sets of estimates of the frequency of from $T = 20$
integrations of the x-axis orbit with $E = -.4059$
in the logarithmic potential (17) of Section 4.1. The
value 2.1390519547 is accurate.

$N = 256$		$N = 512$	
p	ν	p	ν
1	2.1390476260	1	2.1390476324
2	2.1390519936	2	2.1390519918
3	2.1390519526	3	2.1390519541
4	2.1390519586	4	2.1390519547
5	2.1390520877	5	2.1390519547
6	2.1390523943	6	2.1390519547
7	2.1390528801	7	2.1390519547
8	2.1390535210	8	2.1390519547

3.3. ALIASING

Aliasing is an unavoidable consequence of discretization. Two components $e^{i\beta t}$ and
$e^{i\gamma t}$ are indistinguishable on the grid specified in Section 3.1 if $(\beta - \gamma)\Delta t = 2m\pi$
for any integer m. From a practical point of view, two components may interfere
and complicate our method for determining frequencies when $(\beta - \gamma)\Delta t/2\pi$ is
close to an integer. That is the reason for the lack of convergence in the left hand
column of Table I. The aliasing there is between the fundamental frequency ν
and its harmonics with frequencies 37ν and 39ν because $38\nu\Delta t/2\pi = 1.0107$.
(Because x is real, all frequencies occur in $\pm$ pairs.) The G_k for the fundamental
frequency peaks between $k = 13$ and $k = 14$, while those for the two harmonics
peak near $k = 8$ and $k = 19$ respectively. Both of these peaks are much lower
because their amplitudes are $A_{19} = .301 \times 10^{-4}$ and $A_{20} = .235 \times 10^{-4}$, whereas
$A_1 = .8622$. Nevertheless the lower peaks spread with increasing p (see Figure 1)
and cause the improvement of the estimates of ν to stop after $p = 4$. Aliasing
can always be overcome by increasing N and decreasing Δt and moving the peaks
further apart. As the second column of Table I shows, doubling the number of data
points in an interval of the same length removes the difficulty.

3.4. DETERMINING AMPLITUDES

By working with ratios, the amplitude of a Fourier component has no influence on
the estimation of the frequency. However, once one has an accurate estimate of that

frequency, one can estimate its amplitude in $f(t)$, and that of its conjugate $e^{-i\nu t}$, by

$$A_\nu = \frac{F_k^{(p)}}{G_k^{(p)}(\nu)}, \quad A_{-\nu} = \frac{F_\ell^{(p)}}{G_\ell^{(p)}(-\nu)}, \tag{16}$$

respectively.

The iterated Fourier transforms of $f(t) - A_\nu e^{i\nu t} - A_{-\nu}e^{-i\nu t}$, that is of $f(t)$ with the dominant frequency removed, are $F_k^{(p)} - A_\nu G_k^{(p)}(\nu) - A_{-\nu}G_k^{(p)}(-\nu)$. Note that $G_k^{(p)}(-\nu) = G_{-k}^{(p)}(\nu)$. The only additional work needed here is that of calculating the values of $G_k^{(p)}(\nu)$ with the chosen ν for all k in $[-N, N]$. Then one can repeat the procedure iteratively to estimate as many frequencies and amplitudes as needed.

One can modify this step-by-step procedure once the n fundamental frequencies have been determined by assuming that all subsequent frequencies are combinations $\mathbf{k} \cdot \mathbf{\nu}$ suitable to that orbit (see Section 4). These amplitudes can be calculated systematically by applying equation (16) near a peak of $G_k^{(p)}$ and then subtracting. One should subtract in at least approximate order of magnitude. Once one has a set of frequencies and amplitudes, their accuracy can be checked by evaluating the truncated Fourier series and comparing with the tabulated values of f from which it was derived.

4. Applications

We test the DFT method on orbits in the logarithmic potential

$$\Phi(x, y) = \ln\left(R_c^2 + x^2 + \frac{y^2}{q^2} \right), \tag{17}$$

with flattening $q < 1$ and a core radius R_c (Binney and Tremaine 1987). We investigate the same three orbits for the case $R_c = 0.1$ and $q = 0.9$ to which Papaphilippou and Laskar (1996), hereafter PL, applied their NAFF procedures.

4.1. ONE-DIMENSIONAL ORBITS

Though simple dynamically, x-axis orbits for which $y \equiv 0$ provide an excellent opportunity for testing the methods of Section 3 because true values can be computed with high precision. The position vector of an orbit which starts at $x = 0$ at time $t = 0$ has the Fourier representation

$$x(t) = \sum_{j=1}^{\infty} A_j \sin(2j - 1)\nu t. \tag{18}$$

The frequency v for an orbit of energy E is

$$\frac{\pi}{v} = \sqrt{2} \int_0^{x_{max}} \frac{dx}{\sqrt{E - \Phi(x, 0)}}, \qquad \Phi(x_{max}, 0) = E. \tag{19}$$

The Fourier series is a sine series because of the initial conditions, and with odd multiples of vt only because of the evenness of Φ in x. Estimates of the coefficients A_j, which are all positive, can be compared with exact values computed from

$$A_j = \frac{4v}{\pi} \int_0^{\pi/2v} x(t) \sin(2j - 1)vt \, dt. \tag{20}$$

PL give frequencies and amplitudes for the orbit which starts at $x = 0.49$ with $v = 1.4$, and hence $E = -.4059$. The right column of Table I gives our determination its frequency, while Figure 3b plots the Fourier coefficients. After an initial steep decline ($A_2/A_1 = .076$), the rate of decay of A_j with increasing j decreases, and $A_{25}/A_{24} = .798$ at the edge of the plot. This is because x needs many sine terms to describe its time variation accurately, and, correspondingly, the phase-plane trajectory of Figure 3a is not elliptical. The slow decay of the A_j coefficients does not, as PL suggest, indicate any inaccuracy in the representation in action/angle variables. In fact it is related to the small magnitude of the core radius R_c as we now show. The Fourier series (18) of the analytic function $x(t)$ become the Laurent series

$$x = \sum_{j=1}^{\infty} \frac{i}{2} A_j \left[-\zeta^{2j-1} + \frac{1}{\zeta^{2j-1}} \right] \tag{21}$$

in the variable $\zeta = e^{ivt}$ (Davis, 1975). This series converges in an annular region of the complex ζ-plane, which includes the physical orbit on which $|\zeta| = 1$. However, it is singular at the two real values of ζ for which $x = \pm i R_c$ and the logarithmic potential is infinite. We locate these points by integrating the energy equation, written for $x = iw$, from $t = 0$ where $w = 0$ and $\zeta = 1$, to $w = R_c$. We find that the singularity lies at $\zeta = \zeta_c$ where

$$\zeta_c = \exp \left[\frac{v}{\sqrt{2}} \int_0^{R_c} \frac{dw}{\sqrt{E - \ln(R_c^2 - w^2)}} \right]. \tag{22}$$

Integrating in the direction of decreasing ζ shows that there is a matching singularity where $w = -R_c$ at $\zeta = 1/\zeta_c$. Hence the Laurent series (21) converges only in the annulus $1/\zeta_c < |\zeta| < \zeta_c$, which implies (Davis, 1975) that

$$\lim_{j \to \infty} \frac{A_{j+1}}{A_j} = \frac{1}{\zeta_c^2}. \tag{23}$$

The numerical value of this limit for the PL orbit is 0.8702, and this is the limit to which the ratios of the A_j plotted in Figure 3b are gradually climbing.

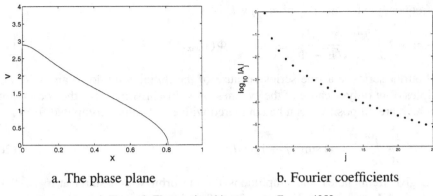

a. The phase plane   b. Fourier coefficients

Figure 3. The x-axis orbit at energy $E = -.4059$

4.2. TWO-DIMENSIONAL ORBITS

The evenness of the potential (17) restricts the two-dimensional Fourier series too. When ν_1 is the fundamental frequency of the oscillation in x, and ν_2 is the frequency of that in y, then the Fourier expansions are

$$x = \sum_{j=-\infty}^{\infty} \sum_{k=-\infty}^{\infty} A_{j,k} e^{i[(2j+1)\nu_1 + 2k\nu_2]t}, \quad A_{-j-1,-k} = \bar{A}_{j,k},$$

$$y = \sum_{j=-\infty}^{\infty} \sum_{k=-\infty}^{\infty} B_{j,k} e^{i[2j\nu_1 + (2k+1)\nu_2]t}, \quad B_{-j,-k-1} = \bar{B}_{j,k}. \tag{24}$$

This is the case for box orbits (Ratcliff, Chang, and Schwarzschild, 1984) while for loop orbits, for which the most prominent motion in both x and y is the circulation about the center, both series have the same form

$$x = \sum_{j=-\infty}^{\infty} \sum_{k=-\infty}^{\infty} A_{j,k} e^{i[(2j+1)\nu_1 + k(\nu_1+\nu_2)]t}, \quad A_{-j-1,-k} = \bar{A}_{j,k},$$

$$y = \sum_{j=-\infty}^{\infty} \sum_{k=-\infty}^{\infty} B_{j,k} e^{i[(2j+1)\nu_1 + k(\nu_1+\nu_2)]t}, \quad B_{-j-1,-k} = \bar{B}_{j,k}. \tag{25}$$

Orbits have another symmetry. If they are started on the x-axis at their maximum excursion in x with initial conditions $x = x_{\max}$, $\dot{x} = 0$, and $y = 0$, then the solutions for x and y are respectively even and odd in t. Hence x has a cosine series and y a sine series. An orbit which is started at some other time will reach $x = x_{\max}$ at time $t = t_0$ say, and so its x and y are even and odd in $(t - t_0)$. Consequently the Fourier coefficients of (24) have the forms

$$A_{j,k} = \pm |A_{j,k}| e^{-i[(2j+1)\nu_1 + 2k\nu_2]t_0}, \quad B_{j,k} = \pm |B_{j,k}| e^{i[\pi/2 - 2j\nu_1 t_0 - (2k+1)\nu_2 t_0]}, \tag{26}$$

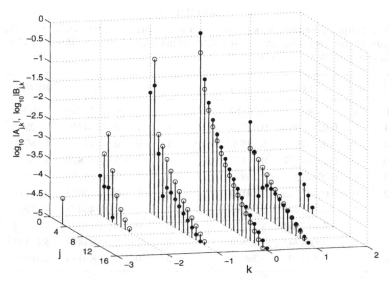

Figure 4. Magnitudes of Fourier coefficients for x, filled circles, and y, open circles, for the box orbit. There are open circles for all $0 \leq j \leq 13$, $k = 1$, but some are hidden by overlapping filled circles.

for some unknown t_0, and similar formulas apply to loop orbits. The arguments of the complex Fourier coefficients allow us to deduce the values of both $\nu_1 t_0$ and $\nu_2 t_0$, though only modulo π.

4.2.1. *A box orbit*

Our estimates of the fundamental x and y frequencies for the box orbit which is at $x = 0.49$, $\dot{x} = 1.3156$, $y = 0$, $\dot{y} = 0.4788$, at $t = 0$ are found from an integration with $T = 100$ and $N = 4096$, as

$$\nu_1 = 2.16322769, \qquad \nu_2 = 3.01399443. \tag{27}$$

There are 53 conjugate pairs of amplitudes $A_{j,k}$ and 51 conjugate pairs of $B_{j,k}$ whose magnitudes exceed 10^{-5}. The ten largest are listed in Table II, and one member of each pair is displayed in Figure 4. Only a few terms in k are needed, but many more in j because the decay in j resembles that in the one-dimensional orbit of Section 4.1. Summing the partial series (24) using the coefficients of magnitude greater than 10^{-5} reproduces the orbit with average errors of 3×10^{-5} at each tabular point.

We can use the Fourier coefficients to calculate the two actions and to verify the Hamiltonian constraints. Equations (4) and (5) require that

$$\sum_{j=-\infty}^{\infty} \sum_{k=-\infty}^{\infty} \left\{ (2j+1) \left[(2j+1)\nu_1 + 2(k-m)\nu_2 \right] A_{j,k} A_{-j-1,m-k} \right.$$
$$\left. + 2j \left[2j\nu_1 + (2k-2m+1)\nu_2 \right] B_{j,k} B_{-j,m-k-1} \right\} = \delta_{mo} I_1,$$

TABLE II

The ten largest Fourier coefficients of the box series (24) for the frequencies (27). The arguments are consistent with equation (26) and the values $\nu_1 t_0 = 1.03468$ (mod π) and $\nu_2 t_0 = 1.00447$ (mod π).

j	k	$|A_{j,k}|$	arg $A_{j,k}$	j	k	$|B_{j,k}|$	arg $B_{j,k}$
0	0	0.38565023	5.24850	0	0	0.12072946	3.70791
1	0	0.02912085	0.03754	1	−1	0.10684730	0.50590
1	−1	0.02252003	5.18808	1	0	0.01611545	4.78014
0	−1	0.01264537	0.97426	2	0	0.00517818	5.85237
2	0	0.00821225	1.10977	3	0	0.00224771	0.64141
3	0	0.00338263	2.18199	2	−2	0.00163069	0.44548
0	1	0.00171443	0.09796	2	−1	0.00140476	4.71972
4	0	0.00167051	3.25422	4	0	0.00113731	1.71363
5	0	0.00091742	4.32644	3	−1	0.00081479	5.79195
6	0	0.00054005	5.39867	5	0	0.00063138	2.78586

$$\sum_{j=-\infty}^{\infty} \sum_{k=-\infty}^{\infty} \left\{ 2k \left[(2j - 2m + 1)\nu_1 + 2k\nu_2 \right] A_{j,k} A_{m-j-1,-k} \right. \tag{28}$$

$$\left. + (2k + 1) \left[2(j - m)\nu_1 + (2k + 1)\nu_2 \right] B_{j,k} B_{m-j,-1-k} \right\} = \delta_{mo} I_2.$$

The constraints are satisfied to an accuracy which is consistent with the truncations of the Fourier series, and the actions are $I_1 = 0.76050$, $I_2 = 0.06422$.

4.2.2. A loop orbit

The two most prominent frequencies of the loop orbit which begin from $x = 0.49$, $\dot{x} = 0.4788$, $y = 0$, $\dot{y} = 1.3156$, found from an integration with $T = 50$ and $N = 2048$, are

$$\nu_1 = 2.948610113, \qquad \nu_2 = 1.357500105. \tag{29}$$

These two fundamental frequencies are related to the frequencies Ω and $\kappa - \Omega$ of epicyclic motion ((Binney and Tremaine, 1987), Chapter 3). They are more easily determined accurately for this orbit which keeps away from the core, and for which the Fourier series converge faster. There are 42 conjugate pairs of amplitudes $A_{j,k}$ and 43 pairs of amplitudes $B_{j,k}$ whose magnitudes exceed 10^{-6}. The ten largest are listed in Table III, and one member of each pair is displayed in Figure 5. Amplitudes decay more rapidly with j than with k. Summing the partial series (25) using the coefficients of magnitude greater than 10^{-6} reproduces the orbit with average errors of 3×10^{-6} at each tabular point.

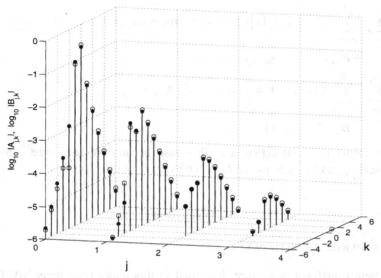

Figure 5. Magnitudes of Fourier coefficients for x, filled circles, and y, open circles, for the loop orbit. Three open circles at $j = 2$ and two at $j = 3$ are hidden by overlapping filled circles.

TABLE III

The ten largest Fourier coefficients of the loop series (25) for the frequencies (29). The arguments are consistent with $\nu_1 t_0 = .33891$ (mod π) and $\nu_2 t_0 = 1.08622$ (mod π) and equation (25).

j	k	$|A_{j,k}|$	arg $A_{j,k}$	j	k	$|B_{j,k}|$	arg $B_{j,k}$
0	0	0.21093349	5.94428	0	0	0.24538048	4.37348
0	−1	0.09038719	1.08622	0	−1	0.07740396	5.79861
0	1	0.01189380	1.37755	0	1	0.01382317	6.08994
1	0	0.00273923	5.26646	1	0	0.00313257	3.69566
1	−2	0.00181571	1.83354	0	2	0.00187210	1.52321
0	2	0.00160147	3.09401	1	−2	0.00137322	0.26275
0	−2	0.00122769	5.65295	1	1	0.00106544	5.41212
1	1	0.00093184	0.69973	1	−1	0.00082623	5.12079
1	−1	0.00089276	0.40841	0	3	0.00033780	3.23967
0	3	0.00028643	4.81047	1	2	0.00032815	0.84539

When equations (4) and (5) are applied to the series (25), we get the conditions

$$\sum_{j=-\infty}^{\infty} \sum_{k=-\infty}^{\infty} (2j+k+1)\left[(2j+k+1)\nu_1 + k\nu_2\right]\left(A_{j,k}A_{-j-1-m,2m-k}\right.$$

$$\left. +B_{j,k}B_{-j-1-m,2m-k}\right) = \delta_{mo}I_1,$$

$$\sum_{j=-\infty}^{\infty} \sum_{k=-\infty}^{\infty} k\left[(2j+k+1)\nu_1 + k\nu_2\right]\left(A_{j,k}A_{m-j-1,-k}\right.$$

$$\left. +B_{j,k}B_{m-j-1,-k}\right) = \delta_{mo}I_2.$$

$$(30)$$

The constraints are well satisfied in this case because of the more convergent Fourier series, and the actions are $I_1 = 0.62879$, $I_2 = 0.04366$.

5. Discussion

5.1. COMPARISONS WITH NAFF

The discrete transform procedure described in this paper resembles NAFF in its basic approach in that it assumes quasi-periodic motion and seeks to calculate an adequate description of that motion. It is more streamlined than existing proced-ures. A single discrete Fourier transform is taken once and there is no need to either calculate Fourier integrals, maximize them, or for orthogonal projection. Test cal-culations show that the procedure attains high accuracy in estimating frequencies and amplitudes, and frequency components do not need to be subtracted twice (Laskar, 1993).

Laskar (1999) proves a theorem which shows that his procedure of locating the maximum of $|\phi(\sigma)|$, where the function $\phi(\sigma)$ is defined as

$$\phi(\sigma) = \frac{1}{2T}\int_{-T}^{T} f(t)\chi_p\left(\frac{t}{T}\right)e^{-i\sigma t}dt,$$

$$(31)$$

locates the most prominent frequency to within $O(1/T^{2p+2})$. This proof assumes that the integral for $\phi(\sigma)$, which has in practice to be evaluated numerically because values of f are known only at tabular points, is known exactly. It is evident that all but one of the orders of T^{-1} in the accuracy come from the windowing, and only the final $(2p+2)$'th from the maximization. Laskar's standard choice is $p = 1$; he reports trying other p values, but not noting much improvement. We do find significant improvements with the use of larger p, and have much success with p in the range of 3 to 5. Our iterative procedure does not lock us into any a priori choice of p, and we can check for convergence and consistency.

We differ from Laskar in that we fit the Fourier series for $\mathbf{x}$, whereas he fits the complex combination $\mathbf{x}+i\mathbf{v}$, which has Fourier coefficients $(1-\mathbf{k}\cdot\boldsymbol{\nu})\mathbf{X_k}$. The series for x has the somewhat more rapidly decaying coefficients which should make it easier to fit. The velocity vector v cannot contain any significant information that is not in x because both are computed simultaneously from accurate numerical integration of an orbit – we use the DOP 853 integrator of Hairer and Wanner (Hairer, Norsett, and Wanner, 1991). Hence we see no reason why fitting $x + iv$ should be any improvement over fitting x.

5.2. Summary and Conclusions

We have shown how the spectral analysis of orbits can be carried out wholly, efficiently and accurately using the DFT. We have given examples, and have investigated the rates of convergence of their Fourier series. This may be slow if the motion in the orbit is far from sinusoidal in time, and an impractically large number of terms may then be needed for an accurate Fourier representation.

There are two choices which have to be made at the outset, the duration $2T$ of the integration and the number $2N$ of outputs. The DOP 853 routine makes it easy to obtain frequent output, and hence to make N sufficiently large to avoid complications from aliasing. The choice of T is significant because the inverse of the DFT (6) is

$$f(t) = \sum_{k=1-N}^{N} F_k e^{\pi i k t / T}. \tag{32}$$

We need T to be sufficiently large that $|\Delta v| T / \pi$ is large for the difference $|\Delta v|$ between any pair of frequencies which are significant in the spectrum, so that they are represented by well-separated k values in expansion (32). We also need $\pi N / T$, the highest frequency in expansion (32) to be large enough that all the frequencies which are present in the true solution are represented in its discrete approximation.

Copin, Zhao, and de Zeeuw (2000) have shown that orbital density is proportional to the reciprocal of the Jacobian of the transformation between position and angles for that orbit. That Jacobian is known from the Fourier series (1), and allows orbital densities, which are needed to construct stellar dynamic models, to be calculated significantly more accurately than is possible with the binning methods (Schwarzschild, 1979) that have been used for so long. The DFT method, which gives a fast and accurate way of determining the Fourier series (1), therefore has great potential for applications to galactic modeling.

Appendix

We show here that the equation

$$G_{k-1}^{(p)}(\theta) = C G_k^{(p)}(\theta), \tag{33}$$

for determining θ, and thence the frequency ν, is a polynomial equation of degree $2p$ in $\tan(\theta/2N)$. It follows from equation (11) that

$$G_k(\theta) = \frac{\sin\theta}{2N\tau}, \quad G_{k+\ell}(\theta) = \frac{(-1)^\ell \sin\theta}{2N} \frac{1+\tau\tau_\ell}{\tau-\tau_\ell}, \tag{34}$$

where

$$\tau = \tan\frac{\theta}{2N}, \quad \tau_\ell = \tan\frac{\ell\pi}{2N}. \tag{35}$$

The iterated coefficient $G_k^{(p)}$ can be related directly to the basic G-coefficients by repeated application of relation (10), or more directly by expanding equation (7) binomially using equation (8), to get

$$\begin{aligned}
G_k^{(p)} &= \frac{(p!)^2}{(2p)!} \sum_{\ell=-p}^{p} \binom{2p}{\ell+p} G_{k+\ell} \\
&= \frac{\sin\theta}{2N\tau} \left[1 + \sum_{\ell=1}^{p} \frac{(-1)^\ell (p!)^2}{(p-\ell)!(p+\ell)!} \frac{2(1+\tau_\ell^2)\tau^2}{\tau^2 - \tau_\ell^2} \right]
\end{aligned} \tag{36}$$

In passing from the first line to the second, we have combined $G_{k\pm\ell}$ terms in pairs. The term in the square brackets is the ratio of two polynomials of degree p in τ^2. Its numerator has $(1+\tau^2)$ as a factor because it vanishes when $\tau^2 = -1$. This is seen by setting $\tau^2 = -1$, and recognizing the sum as the binomial expansion of $(p!)^2(1-1)^{2p}/(2p)! = 0$.

The other iterated coefficient of

$$G_{k-1}^{(p)} = \frac{(p!)^2}{(2p)!} \sum_{\ell=-p}^{p} \binom{2p}{\ell+p} G_{k-1+\ell}, \tag{37}$$

is less symmetric in τ, but it is likewise rational in τ with a numerator polynomial of degree $2p$ divided by a denominator polynomial of degree $(2p+1)$. Its denominator differs from that of $G_k^{(p)}$ only in having a factor $(\tau - \tau_{-p-1})$ and lacking one in $(\tau - \tau_p)$. Furthermore the numerator of $G_{k-1}^{(p)}$ also has $(1+\tau^2)$ as a factor. This follows from the alternative and symmetric expression for it we get from equation (36) when we replace τ by $\tilde{\tau} = \tan[(\theta+\pi)/2N] = (\tau+\tau_1)/(1-\tau\tau_1)$, because $(1+\tilde{\tau}^2) = (1+\tau^2)(1+\tau_1^2)/(1-\tau\tau_1)^2$. Consequently when we multiply equation (33) by all denominator terms and cancel all common factors, a polynomial equation of degree $2p$ in τ remains.

Acknowledgements

It is a pleasure to thank Alex Fridman for organizing a stimulating workshop, and him and the Director and staff of the Sternberg Astronomical Institute for their hospitality. This work has been supported in part by the National Science Foundation through grant DMS-9704615.

References

Binney, J. and Spergel, D.: 1982, 'Spectral stellar dynamics', *Astrophys. J.* **252**, 308.
Binney, J. and Spergel, D.: 1984, 'Spectral stellar dynamics – II. The action integrals', *Monthly Notices Royal Astron. Soc.* **206**, 159.
Binney, J. and Tremaine, S.: 1987, *Galactic Dynamics*. Princeton Univ. Press, Princeton, NJ, USA.
Briggs, W. L. and Henson, V. E.: 1995, 'The DFT: an owner's manual of the discrete Fourier transform', Society for Industrial and Applied Mathematics, Philadelphia, PA, USA.
Carpintero, D. D. and Aguilar, L. A.: 1998, 'Orbit classification in arbitrary 2D and 3D potentials', *Monthly Notices Royal Astron. Soc.* **298**, 1.
Copin, Y., Zhao, H. S. and de Zeeuw, P. T.: 2000, 'Probing a regular orbit with spectral dynamics', *Monthly Notices Royal Astron. Soc.* **318**, 781.
Davis, P. J.: 1975, 'Interpolation and Approximation', Dover, New York, NY, USA.
Hairer, E., Norsett, S. P. and Wanner, G.: 1991, 'Solving Ordinary Differential Equations I', Springer Verlag, New York, NY, USA.
Lanczos, C.: 1956, 'Applied Analysis', Prentice Hall, Englewood Cliffs, NJ, USA.
Laskar, J.: 1990, 'The chaotic motion of the solar system: a numerical estimate of the size of the chaotic zones', *Icarus* **88**, 266.
Laskar, J.: 1993, 'Frequency analysis for multi-dimensional systems. Global dynamics and diffusion', *Physica D* **67**, 257.
Laskar, J.: 1999, 'Introduction to frequency map analysis', in Siwo, C. (ed.), *Hamiltonian Systems with Three or more Degrees of Freedom.* Kluwer, Dordrecht, Netherlands.
Merritt, D. and Valluri, M.: 1999, 'Resonant orbits in triaxial galaxies', *Astron. J.* **118**, 1177.
Papaphilippou, Y. and Laskar, J.: 1996, 'Frequency map analysis and global dynamics in a galactic potential with two degrees of freedom', *Astron. Astrophys.* **307**, 427.
Papaphilippou, Y. and Laskar, J.: 1998, 'Global dynamics of triaxial galactic models through frequency map analysis', *Astron. Astrophys.* **329**, 451.
Ratcliff, S. J., Chang, K. M. and Schwarzschild, M.: 1984, 'Stellar orbits in angle variables', *Astrophys. J.* **279**, 610.
Schwarzschild, M.: 1979, 'A numerical model for a triaxial stellar system in dynamical equilibrium', *Astrophys. J.* **232**, 236.
Valluri, M. and Merritt, D.: 1998, 'Regular and chaotic dynamics of triaxial stellar systems', *Astrophys. J.* **506**, 686.
Wachlin, F. C. and Ferraz-Mello, S.: 1998, 'Frequency map analysis of the orbital structure in elliptical galaxies', *Monthly Notices Royal Astron. Soc.* **298**, 22.

SHOULD ELLIPTICAL GALAXIES BE IDEALISED AS COLLISIONLESS EQUILIBRIA?

HENRY E. KANDRUP

Department of Astronomy, Department of Physics, and Institute for Fundamental Theory, University of Florida, Gainesville, Florida, USA 32611 (E-mail: kandrup@astro.ufl.edu)

Abstract. This review summarises several different lines of argument suggesting that one should not expect cuspy nonaxisymmetric galaxies to exist as robust, long-lived collisionless equilibria, *i.e.*, that such objects should not be idealised as time-independent solutions to the collisionless Boltzmann equation.

1. Motivation

Should elliptical galaxies be visualised as objects that spend most of their lives in or near a true equilibrium, continually disturbed somewhat by nearby objects, but not exhibiting any systematic dynamical evolution? Or should they be viewed instead as objects that may be close to some equilibrium, but are drifting through phase space in such a fashion as to manifest a systematic secular evolution on time scales shorter than t_H, the age of the Universe? In other words: once they have settled down towards a near-equilibrium, should galaxies be viewed as static or dynamic entities?

Over the age of the Universe, ellipticals clearly change in terms of such properties as colour. However, it is probably fair to say that, until recently, many, if not most, astronomers have typically ignored the possibility of systematic dynamical changes except, perhaps, in response to very close galaxy-galaxy encounters, arguing, e.g., that discreteness effects (i.e., gravitational Rutherford scattering between individual stars) should be unimportant since the natural relaxation time $t_R \gg t_H$. The object of this review is to argue that this conventional wisdom may not be completely correct. Complex equilibria, especially cuspy and/or nonaxisymmetric equilibria, may not exist or may be very hard for a real galaxy to find; and, especially in high density environments, external irregularities may be significantly more important in triggering systematic changes in the bulk structure of a galaxy than has been generally recognised hitherto.

2. Can Realistic Equilibrium Models Exist?

Dating back to the pioneering work of Eddington and Jeans, galactic dynamicists have typically tried to construct equilibrium solutions to the collisionless Boltzmann equation (*CBE*) using global integrals like the energy E and rotational angular momentum J_z. Since these quantities are conserved, any function of them will also be conserved, so that they can be used to define an equilibrium phase space density, or distribution function, f_0. However, such 'standard' equilibria are problematic if one allows for the possibility that many galaxies are at least moderately triaxial, as one seems compelled to do from an analysis of observations (cf. Tremblay and Merritt, 1995) Generic time-independent, albeit nonaxisymmetric, potentials admit only one continuous symmetry, namely time-translation, so that there is only one global integral. If the system is nonrotating, this corresponds to the energy E; if, instead, the system is in uniform rotation, this corresponds to the Jacobi integral E_J.

The problem, then, is that it seems impossible to model triaxial configurations with a strong central condensation in terms of equilibrium solutions $f_0(E)$ or $f_0(E_J)$. One knows, e.g., that every $f_0(E)$ must correspond to a spherically symmetric configuration (Perez and Aly, 1996). A rotating solution $f_0(E_J)$ *can* in principle correspond to a triaxial configuration, but such configurations seem unrealistic (cf. Vandervoort, 1980; Ipser and Managan, 1981). Physically these triaxial configurations can exist for the same reason as the triaxial Riemann ellipsoids, namely because it is energetically favourable for the configuration to deviate from axisymmetry. However, this requires rotation rates that are too large to be realistic and central condensations much smaller than those observed in real early-type galaxies (cf. Lauer *et al.*, 1995). For example, triaxial equilibria with distribution functions appropriate for a rotating polytrope cannot have a central density larger than about 3.12 times the mean density.

One can of course seek to construct models with 'unobvious' symmetries such as, e.g., models corresponding to integrable Stäckel potentials. However, nonaxisymmetric models admitting two or three global integrals are extremely nongeneric, possessing continuous symmetries whose physical origins are far from obvious. Three-integral potentials constitute a set of measure zero in the set of genuinely three-dimensional potentials and, as such, if one wants to model real galaxies with such integrable potentials, it would seem crucial to identify a guiding principle which would explain why it is that nature selects such equilibria. One might perhaps hope to model triaxial galaxies in terms of potentials that admit only two global integrals. However, it is not clear whether there exist realistic potentials that admit only two global integrals which are not axisymmetric; and, even presuming that such potentials exist, these would again seem comparatively nongeneric.

Alternatively, one can look for equilibria involving 'local integrals' (cf. Lichtenberg and Lieberman, 1992), the conserved quantities that make regular orbits behave as if they were integrable, even if the potential is nonintegrable and chaotic orbits also exist (Kandrup, 1998a). One can, e.g., try to construct equilibria which contain both regular and chaotic orbits but which, for fixed E or E_J, assign different weights to the regular and chaotic phase space regions. This is, in fact, the tack implicit in almost all work with nonintegrable potentials that break axisymmetry, including the triaxial generalisations of the Dehnen (1993) potentials considered, e.g., by Merritt and Fridman (1996).

This problem can be addressed numerically using some variant of Schwarzschild's (1979) method. What this entails is (i) specifying the presumed mass density ρ and gravitational potential Φ for some time-independent equilibrium, (ii) generating a huge library of orbits evolved in this potential, and then (iii) trying to select a weighted ensemble of orbits from that library which reproduces the assumed density. Only by demanding that this weighted ensemble reproduce the original ρ can one ensure that one has a true self-consistent equilibrium, in which the matter in the galaxy evolves in the potential generated by the matter itself.

The usual implementation of this method might seem problematic in that it ignores the role of conserved quantities like energy. However, it was shown some years ago (Vandervoort, 1984) that these quantities are really hiding in the method, at least for the special case of integrable potentials, and more recently Kandrup (1998a) has shown that this method can also be modified in a natural fashion to incorporate 'local integrals'. The key observation is that it is not orbits *per se* that should be considered as the fundamental ingredients. Rather, the crucial point is to select a collection of time-independent building blocks which, being time-independent, can be used as static constituents for a time-independent equilibrium. Very recently, Copin, Zhao, and de Zeeuw (2000) have shown how, for the special case of an integrable potential, one can actually proceed semi-analytically, provided only that action-angle variables can be implemented explicitly: There is a one-to-one correspondence between values of the actions and time-independent density building blocks; and, by sweeping through all possible values of the actions, one is guaranteed to consider all possible time-independent building blocks. However, it does not seem possible to generalise this approach to nonintegrable potentials.

Unfortunately, there are obvious problems with Schwarzschild's method or any other numerical scheme. In particular, there is no proof that, for any given ρ, a self-consistent equilibrium exists or that that equilibrium is unique. Indeed, even assuming that exact solutions do exist, there is no guarantee that a numerical 'solution' is a reasonable approximation to some real solution. Whether or not an equilibrium exists, the numerical prescription will find some 'best fit' solution, and there is no reason *a priori* to assume that that 'best fit' corresponds to a *bona fide* collisionless equilibrium. Alternatively, the inability to generate anything remotely resembling a 'reasonable' solution need not guarantee that no solution exists: this may simply signal an incomplete orbital library. At the present time, the best that one might hope to do is sample one's purported equilibria to generate N-body realisations, evolve these N-body realisations into the future, and then determine whether these behave more or less stably for a finite time. However, even this is problematic. The normal discretisation involved in implementing Schwarzschild's method is so coarse that N-body realisations generated from Schwarzschild models of stable equilibria like Plummer spheres can behave unstably Siopis (2001).

Another, potentially even more serious, problem with almost all work hitherto is that the density distributions which have been considered are highly idealised. Most Schwarzschild modeling has involved an assumption of strict axisymmetry or strict ellipsoidal symmetry with constant axis ratios. For example, the claim that triaxial equilibria cannot exist for galaxies with very steep cusps is based almost completely on an analysis of the triaxial generalisations of the Dehnen potentials (Merritt and Fridman, 1996). However, it is by now well established that real ellipticals tend to have distinctly disky or boxy isophotes, the details of which correlate with other properties of the galaxy, such as the steepness of the central density cusp or the bulk rotation rate (cf. Kormendy and Bender, 1996). Moreover, even if a galaxy is nearly axisymmetric in the center, as seems likely for the coreless ellipticals, they could be distinctly non-axisymmetric in their outer regions. The claim that triaxial potentials containing a very large supermassive black hole have 'too many' chaotic orbits (cf. Merritt and Valluri, 1999) may well reflect unnatural attempts to combine potentials with incompatible symmetries (cf. Kandrup and Sideris, 2001): the black hole is presumably nearly spherical or axisymmetric, but one typically assumes that the surrounding galaxy is triaxial with fixed axis ratios down to very small scales.

In any event, the fact that real galaxies are disky or boxy is probably no accident, and it would seem that realistic galactic models should become more nearly axisymmetric towards the center. However, such 'more complex' objects might seem even less likely to manifest continuous symmetries that give rise to global integrals, so that, assuming that they exist, equilibria with such shapes would be even more likely to rely on 'local integrals'.

3. Will Real Galaxies Evolve Towards Such Equilibria?

Over the past several decades, considerable effort, both analytic and numerical, has been devoted to the construction of equilibrium solutions to the collisionless Boltzmann equation. Unfortunately, however, much less is known about a time-dependent evolution governed by the *CBE*. There is, e.g., no proof that generic initial data will evolve towards a time-independent equilibrium, even assuming that the configuration is gravitationally bound. Even such a basic property as *global existence*, i.e., the fact that $f(t)$ does not evolve singularities and/or caustics, was only proven in the early 1990's (Pfaffelmoser, 1992; Schaeffer, 1991). It would seem that the only hard results about a time-dependent evolution that have been established to date concern the behaviour of quantities like time-averaged moments in an asymptotic $t \to \infty$ limit (cf. Batt, 1987). That so little is known is not really surprising. Even for the seemingly simple case of mechanical systems with short range forces, it is often very difficult, if not impossible, to prove that there is any approach towards equilibrium (cf. Sinai, 2000).

In this regard, it should be emphasised that there exist exact time-dependent solutions to the *CBE* which do *not* manifest any approach towards an equilibrium. For example, Louis and Gerhard (1988) used semi-analytic techniques to construct a solution which corresponds to finite amplitude, undamped oscillations about an otherwise time-independent equilibrium f_0. Moreover, at least for the toy model of one-dimensional gravity, counter-streaming initial conditions, corresponding to a collision between two galaxies initially in equilibrium, can yield a numerical evolution towards a final state with undamped oscillations (cf. Mineau, Feix, and Rouet, 1990). *A priori* it might seem surprising that such oscillations do not exhibit linear and/or nonlinear Landau damping which would cause them to phase mix away. The crucial point physically is that these solutions contain 'phase space holes,' i.e., regions in the middle of the otherwise occupied phase space regions where $f \to 0$, so that one has the possibility of excitations that do not undergo a particle-wave interaction.

The idea that oscillations never damp may seem too extreme to be realistic. However these results would suggest that internal irregularities could persist much longer than the theorist is wont to assume.

In any event, because an evolution governed by the *CBE* is Hamiltonian (cf. Kandrup, 1998c), any statement regarding an approach towards equilibrium must entail a coarse-grained description of the system. This could, e.g., involve a consideration of coarse-grained distribution functions, as in the original proposal of violent relaxation (Lynden-Bell, 1967). Alternatively, one might consider the evolution of a collection of lower order moments, e.g., in the context of a cumulant expansion, an approach well known from plasma physics and accelerator dynamics. In either setting, the obvious question is: how fast, and how completely, do the observables of interest – either coarse-grained distributions or lower order moments – approach time-independent values?

In a galaxy that is far from equilibrium, it is not unlikely that many orbits will exhibit an exponentially sensitive dependence on initial conditions, and the resulting *chaotic mixing* (cf. Kandrup and Mahon, 1994; Merritt and Valluri, 1996; Kandrup, 1998b) will certainly help a galaxy shuffle itself up. However, there is no guarantee that the galaxy will settle down all that efficiently towards a time-independent or nearly time-independent equilibrium! Indeed, recent numerical simulations (cf. Vesperini and Weinberg, 2000) suggest that close encounters between galaxies can lead to long-lived pulsations. (See also the oscillations described in Miller and Smith, 1994.) It will be argued below that such oscillations could in fact trigger secular evolution on a time scale $< t_H$.

Analyses of flows in a fixed potential suggest (cf. Kandrup, 1998b) that, when evolved into the future, generic ensembles of initial conditions *do* eventually exhibit a coarse-grained approach towards equilibrium. However, this can be an extremely complex, multi-stage process for nonintegrable potentials with a phase space that admits a complex coexistence of both regular and chaotic regions and, consequently, is riddled by a complex Arnold web. In particular, it is apparent that the rate of approach towards equilibrium can depend sensitively on the level of coarse-graining: probing the system at different scales and/or in terms of different order moments can lead to significant differences in the rate associated with any approach towards equilibrium. This is hardly surprising given the physical expectation that, as a result of phase mixing, power should cascade from larger to shorter scales.

The obvious point in all this is that great care must be taken in deciding what one ought to mean by asserting that a galaxy is 'nearly in equilibrium.' A galaxy could, e.g., look 'nearly in equilibrium' when viewed in terms of its bulk properties, but still exhibit significant shorter scale variability and, most importantly, still be distinctly 'out of equilibrium' from the standpoint of dynamics.

A crucial point, then, is that systems that have not achieved a true time-independent equilibrium tend to be more susceptible to external stimuli than systems that are in a true equilibrium. This is especially true for complex systems characterised by a six-dimensional phase space that admits a complex coexistence of regular and chaotic regions, with structures like bars or cusps that rely on an intricate balance of 'sticky' (cf. Contopoulos, 1971) and wildly chaotic orbits.

Real galaxies, of course, are not strictly collisionless, and it is clear that dissipative gas dynamics must have played an important role in the earliest stages of galaxy formation where, even allowing for large quantities of nonbaryonic dark matter, much of the (proto-)galaxy is comprised of dissipative gas which has not yet been converted into stars. The obvious point, then, is that dissipative effects associated with this gas could play an important role in driving the system towards a true equilibrium. Indeed, such dissipative effects, which should be more important in the higher density central regions, might drive the central regions of a cuspy galaxy towards a state that is much more nearly axisymmetric than the lower density, outer portions of the galaxy.

Nevertheless, even if dissipation is a crucial element for the evolution of primordial equilibria or near-equilibria, it would seem comparatively unimportant when considering the effects of recent collisions and other close encounters between ellipticals. Over the past decade or so, it has been recognised that ellipticals are not as gas-poor as was originally believed. However, most of the gas in ellipticals exists at high temperatures and low densities, so that its dissipative effects should be minimal. It seems implausible that dissipative gas dynamics could play a dominant role in how elliptical galaxies readjust themselves after a strong encounter with another galaxy.

However, dissipative gas dynamics is not the only physical effect which is ignored by the *CBE*. Real galaxies are also subjected to a variety of other perturbations which could in principle be important. For example, as stressed, e.g., by Merritt and Fridman (1996), the central regions of cuspy galaxies are so dense that the relaxation time t_R associated with gravitational Rutherford scattering between neighbouring stars can be less than or comparable to t_H. And similarly, galaxies are subjected continually to perturbations reflecting the effects of companion galaxies and other nearby objects which, especially in high density clusters, can be appreciable.

An obvious question, therefore, is: what are the potential effects of ongoing low amplitude perturbations? On the one hand, one might argue that they will interfere with a systematic approach towards equilibrium since they imply that the density distribution necessarily varies in time. On the other, one might argue that these perturbations actually expedite the approach towards equilibrium, since the time-varying forces to which the galaxy is subjected could facilitate violent relaxation by accelerating phase space transport. In particular, it would not seem completely implausible to argue that, in many cases, galactic evolution should be viewed as a two-stage process. Early on a galaxy could have evolved towards a state which, albeit not a true equilibrium, would persist as a near-equilibrium for times $> t_H$ in the absence of any irregularities. However, such irregularities are always present, and they could act to trigger a secular evolution, e.g., driving the system more nearly towards a true equilibrium.

4. What are the Effects of Ongoing Perturbations?

Suppose that, after some more or less effective period of violent relaxation, an elliptical has settled down *towards* (albeit not necessarily *to*) some realistic complex equilibrium or near-equilibrium. That configuration could well be distinctly non-axisymmetric, at least in the outer regions, and, if so, might be expected to exhibit variable axis ratios, becoming more nearly axisymmetric near the center. However, the gravitational potential associated with such a complex configuration is likely to involve significant measures of both regular and chaotic orbits. It is in fact well known to nonlinear dynamicists that less symmetric potentials tend to exhibit much larger measures of chaos, a point first stressed in the galactic dynamics community by Udry and Pfenniger (1988). Even comparatively simple potentials like the three-dimensional logarithmic potentials, with $V = \frac{1}{2}v_0^2 \log(1 + x^2/a^2 + y^2/b^2 + z^2/c^2)$, admit chaotic orbits for certain choices of parameter values. Indeed, there exist self-consistent axisymmetric equilibria which manifest chaotic meridional motions, including the scale-free models considered by Evans (1994) and various spheroidal models considered by Hunter *et al.* (1998).

The important point, then, is that the phase space associated with a time-independent three-dimensional potential that admits significant measures of both regular and chaotic orbits tends to be very complex, being laced with cantori and/or an Arnold web. The existence of this complex structure implies that the approach towards equilibrium can be comparatively inefficient. In particular, orbits could well get trapped in localised phase space regions for very long times, even though their motion is not blocked by a conservation law (like conservation of energy): In principle, an orbit may be able to access a large phase space region, but it may have to leak through a narrow 'bottleneck' to get from one part of the accessible region to another, penetrating through what the nonlinear dynamicist would call an *entropy barrier.*

This suggests the possibility that a system could evolve towards a state which, albeit not a true equilibrium, could persist for times $\gg t_H$, at least in the absence of significant perturbations. Indeed, this possibility has been invoked by a number of authors (cf. Patsis, Athanassoula, and Quillen, 1997) in the context of spiral galaxies, and corresponds to what Merritt and Fridman (1996) termed 'quasi-equilibria' involving chaotic building blocks that are only 'partially mixed'.

The important observation, however, is that such configurations can be surprisingly vulnerable to even very weak irregularities. This fact was apparently first recognised more than thirty years ago in the context of simple maps (cf. Lieberman and Lichtenberg, 1970); and, over the past several decades has proved to be very important in the context of plasma physics and accelerator dynamics (cf. Tennyson, 1979; Habib and Ryne, 1995), where one deals with beams of charged particles that are confined by imperfect magnetic fields and in which discreteness effects, i.e., Rutherford scattering, can be surprisingly important.

But what can low amplitude perturbations actually do? Even if the perturbations are too weak to significantly impact the values of the energy or any other collisionless invariants, they can induce systematic phase space flows on the (nearly) constant energy surfaces, e.g., by allowing orbits originally trapped in localised phase space regions to become untrapped (cf. Mahon *et al.*, 1995; Kandrup, Pogorelov, and Sideris, 2000) Moreover, under appropriate circumstances, perturbations can prove strong enough to induce nontrivial changes in the orbital energy and other collisionless invariants (cf. Kandrup, 2001).

Such perturbations act via a resonant coupling between the frequencies of the perturbation and the frequencies of the orbits (Pogorelov and Kandrup, 1999). For regular orbits, the power is concentrated at a few special frequencies whereas, for chaotic orbits, power is typically broader band. In either case, however, when subjected to a perturbation characterised by its own set of natural frequencies, there is the possibility of resonance; and, to the extent that this resonance is strong, the perturbation will have a substantial effect.

One might suppose that the response of the orbits to various influences will depend sensitively on the precise form of the perturbation, so that believeable computations would require a knowledge of details that are difficult, if not impossible, to extract from observations. In point of fact, however, the details seem comparatively unimportant. Because the perturbations act via a resonant coupling, all that really seems to matter are the amplitude of the perturbation, which determines how hard the orbits are being 'hit,' and the natural frequencies of the perturbation, which determine the degree to which a resonance is possible.

The crucial point about this susceptibility to low amplitude irregularities is that changes in the orbital density and/or collisionless invariants will in general yield changes in the bulk potential and, as such, could trigger systematic evolutionary effects. It could be that the system will react in such a fashion as to stabilise itself, but it might equally well start to exhibit systematic changes in its bulk properties.

The most detailed investigation of the effects of low amplitude perturbations in the context of elliptical galaxies that has been effected hitherto (Siopis and Kandrup, 2000; Kandrup and Siopis, 2002) involved an investigation of flows in the triaxial generalisations of the Dehnen potentials, with or without a central supermassive black hole. These potentials are not completely realistic since they assume strict ellipsoidal symmetry with fixed axis ratios, and ignore completely the possibility of rotation. However, they do at least incorporate a high density central concentration and, as might be expected physically, they include large measures of both regular and chaotic orbits. Several different physical effects were considered:

- *Discreteness effects*, i.e., gravitational Rutherford scattering between individual stars, were modeled as resulting in dynamical friction and Gaussian *white noise*, i.e., near-instantaneous impulses, in the spirit of Chandrasekhar (1943). Even if the relaxation time t_R on which the energy of individual orbits changes is long compared with t_H which, as stressed by Merritt and Fridman (1996), is not always so, such encounters can still be important by accelerating diffusion on the nearly constant energy hypersurface (cf. Habib, Kandrup, and Mahon, 1997; Kandrup, Pogorelov, and Sideris, 2000).

- The effects of *satellite galaxies and companion objects* were modeled as *near-periodic driving*, characterised by a perturbation $V = V(\{\omega_i\}t)$, for some small number of frequencies $\{\omega_i\}$.

- In *a dense cluster environment*, a galaxy will be impacted by a large number of different neighbouring galaxies in a fashion which is likely to be far from periodic. It thus seemed reasonable to model such an environment by allowing for *coloured noise*, this corresponding (cf. van Kampen, 1981) to a series of random impulses of finite duration. (Mathematically, coloured noise can be viewed as a superposition of periodic disturbances with different frequencies combined with random phases.)

- In addition to companion objects, a galaxy can be impacted by *tidal forces* associated with the bulk cluster potential which, depending on the form of the potential, could be *nearly periodic* or *largely random*.

- *Coherent internal oscillations*, associated with a small number of normal or pseudo-normal modes, were again modeled as inducing a *near-periodic driving*.

- *Incoherent internal oscillations*, associated, e.g., with a large number of higher order modes, were again modeled as *coloured noise*, i.e., near-random perturbations of finite duration. (Making a sharp distinction between coherent and incoherent oscillations is admittedly somewhat *ad hoc*.)

Given the usual assumption that the noise under consideration is Gaussian, its statistical properties are characterised completely by a knowledge of its first two moments. For both white and coloured noise, it was assumed that the average force vanishes identically, so that $\langle F_a(t) \rangle = 0$ for $a = x, y, z$. White noise, which involves instantaneous kicks, is characterised by an autocorrelation function

$$\langle F_a(t_1) F_b(t_2) \rangle = 2D \, \delta_{ab} \delta_D(t_1 - t_2). \tag{1}$$

Coloured noise requires 'fuzzing out' the Dirac delta. As a simple example, this was done by sampling an Ornstein-Uhlenbeck process, for which

$$\langle F_a(t_1) F_b(t_2) \rangle = (D/t_c)\delta_{ab} \, \exp(-|t_1 - t_2|/t_c). \tag{2}$$

In each of these expressions, D is the diffusion constant which, in the white noise limit, enters into a Fokker-Planck description:

$$D \equiv \int_0^\infty d\tau \, \langle F_a(0) F_a(\tau) \rangle. \tag{3}$$

The quantity t_c is the autocorrelation time, which represents the characteristic time scale on which the random forces change significantly.

The effects of low amplitude perturbations can be decomposed, at least approximately, into (i) motions that involve little or no changes in the energy and any other quantities which would be invariant for motion in a fixed time-independent potential; and (ii) changes in the values of the energy and any other collisionless invariants.

Even if the perturbations are so weak that such collisionless invariants as E or E_J are nearly conserved, they can significantly accelerate phase space diffusion of chaotic orbits on the nearly constant energy surface. What this entails is orbits leaking through topological partial obstructions associated with cantori and/or an Arnold web in a fashion strongly reminiscent of the standard problem of *effusion*, whereby gas molecules can leak through a tiny hole in a wall. If one considers an ensemble of 'sticky' orbits (cf. Contopoulos, 1971) initially localised in a given phase space region bounded by such obstructions, their escape is well modelled as a Poisson process (cf. Pogorelov and Kandrup, 1999; Kandrup, Pogorelov, and Sideris, 2000), for which the number of orbits that remain trapped decreases exponentially, i.e.,

$$N(t) = N_0 \exp(-\Lambda t). \tag{4}$$

For a specified choice for the form of the perturbation, the value of Λ scales logarithmically in amplitude, so that, e.g., for the case of noise there is a logarithmic dependence on D. The effects tend to be significant provided only that the characteristic time scale associated with the perturbation is not large compared with the natural time scale associated with the orbits. For the case of coloured noise, the precise choice of autocorrelation time seems immaterial for $t_c < t_D$, but the effects of the perturbation diminish significantly for $t_c \gg t_D$.

Under appropriate circumstances, the perturbations can also induce nontrivial changes in such collisionless invariants as the energy. For the case of periodic driving, where the perturbations are characterised by a few special frequencies, the detailed effects of the perturbation can be comparatively complex (cf. Kandrup, Abernathy, and Bradley, 1995). However, for the case of 'random' disturbances, idealised as noise, the picture is quite simple, with the perturbations acting to trigger a *diffusion process*. For the case of white or near-white noise with $t_c \ll t_D$, the physics is identical to Brownian motion, as proposed originally by Chandrasekhar (1943) to model the effects of gravitational Rutherford scattering between individual stars or by Spitzer and Schwarzschild (1951) to account for the scattering of disc stars off of giant molecular clouds. For $t_c > t_D$ the effects once again begin to be suppressed. For the special case of coloured noise sampling the Ornstein-Uhlenbeck process, the *rms* change in energy is in general well fit by a scaling relation of the form (Kandrup, 2001)

$$\frac{\delta E_{rms}}{|E|} \sim \left(\frac{Dt}{|E|}\right)^{1/2} \times \begin{cases} 1 & \text{for } t_c < t_D \\ \left(\frac{t_D}{t_c}\right) & \text{for } t_c > t_D. \end{cases} \tag{5}$$

Overall, regular and chaotic orbits are affected in a nearly identical fashion, except for very large values of t_c, where regular orbits prove to be somewhat *more* susceptible than are chaotic orbits.

The important point in all this is that realistic choices of D and t_c can actually have a significant effect. The diffusion constant scales as $D \sim F^2 t_c$, where F denotes the typical size of the random forces, so that its amplitude is readily estimated.

On the nearly constant energy hypersurface, discreteness effects corresponding to a relaxation time as long as $t_R \sim 10^6 - 10^7 t_D$ or even more can have big effects within a time as short as $100\, t_D$. Similarly, even ignoring the effects of single very close encounters, random interactions with nearby galaxies in an environment where the separation between galaxies is ten times the size of an individual galaxy, i.e., $R_{sep} \sim 10\, R_{gal}$, so that the autocorrelation time $t_c \sim 10\, t_D$, can have appreciable effects within $100\, t_D$ (Siopis and Kandrup, 2000; Kandrup and Siopis, 2002).

As regards changes in the energy and other collisionless invariants (Kandrup, 2002; Kandrup and Siopis, 2001), incoherent internal oscillations at the 1% level, far too small to be detected observationally, can trigger 10% changes in the energy within $100\, t_D$; and 10% oscillations can trigger 30% changes in a comparable time. Random interactions between neighbouring galaxies are less important, at least directly, since an environment with $R_{sep} \sim 6\, R_{gal}$ and $t_c \sim 6\, t_D$ will only trigger 10% changes within $100 t_D$. However, such interactions could still prove important (cf. Vesperini and Weinberg, 2000) by triggering random incoherent oscillations that induce more substantial changes in the invariants.

5. Discussion

A major issue all too often unaddressed is the extent to which, as has been assumed here, real galaxies characterised by the highly irregular potential associated with a large number of nearly point mass stars can in fact be approximated by a smooth three-dimensional potential. Should one, for example, really expect to see effects like 'stickiness' or phase space diffusion, which have been studied primarily in the context of smooth two- and three-dimensional potentials, in real many-body systems? Indeed, it is clear that the continuum approximation misses at least some physical effects. It has, for example, become evident from both numerical computations (cf. Goodman, Heggie, and Hut, 1993) and rigorous analytics (Pogorelov, 2001) that, even for very large N, individual orbits in an N-body system typically have large positive Lyapunov exponents, even if the N-body system corresponds to an integrable density distribution such as a spherical equilibrium.

However, recent numerical work involving orbits and orbit ensembles evolved in fixed N-body realisations of continuous density distributions has shown that, in many respects, such 'frozen-N' orbits are indistinguishable from orbits evolved in the continuous density distribution in the presence of friction and (nearly) white noise (Kandrup and Sideris, 2001; Sideris and Kandrup, 2002).

Even though the Lyapunov exponents for frozen-N orbits do not appear to converge towards the values assumed by orbits in the potential associated with the smooth density distribution, there is a precise sense in which, as N increases, frozen-N trajectories remain 'close to' smooth potential characteristics with the same initial conditions for progressively longer times. Viewed macroscopically, for both regular and chaotic initial conditions, frozen-N trajectories and smooth potential characteristics with the same initial condition typically exhibit a *linear* divergence: Their mean separation $\delta r(t)$ is well fit by a growth law $\delta r/R = A(t/t_G)$, where R represents the size of the configuration space region accessible to the orbits and A is a constant of order unity. For regular orbits, $t_G/t_D \propto N^{1/2}$; for chaotic orbits, $t_G/t_D \propto \ln N$. Moreover, for sufficiently large N ensembles of frozen-N orbits can exhibit 'stickiness' qualitatively similar to what has been observed for orbits in smooth potentials. And finally, both in terms of the statistical properties of orbit ensembles *and* in terms of the pointwise properties of individual orbits, discreteness effects associated with a frozen-N system can be well mimicked by white noise with a diffusion constant exhibiting (at least approximately) the N-dependence predicted when these effects are modeled as a sequence of incoherent binary encounters (cf. Chandrasekhar, 1943).

The extent to which the smooth potential approximation remains valid in the context of a self-consistent N-body evolution is much more difficult to probe directly. However, recent work in the context of charged accelerator beams, where particles interacting via a repulsive $1/r^2$ force are confined by an externally imposed potential, appears encouraging. In particular, fully self-consistent grid code simulations of intensed charged beams (which admittedly suppress the chaos associated with close encounters between individual particles) have been found to exhibit both regular and chaotic phase mixing qualitatively similar to what has been observed in smooth two- and three-dimensional potentials (Kishek *et al.*, 2001). Experiments to search for these effects in real particle beams are currently in the planning stage (Bohn *et al.*, 2002) and analogous computations for self-gravitating systems are underway.

The principal message of this review is that it may be oversimplistic to assume that elliptical galaxies should be viewed as collisionless equilibria. Realistic equilibria, corresponding to nonaxisymmetric systems which contain a high density central region, break strict ellipsoidal symmetry, and manifest variable axis ratios, may not exist and, even if they do exist, may be very difficult for real galaxies to find. Moreover, even if a galaxy seems 'close to' equilibrium observationally, it could well be comparatively 'far from' equilibrium from the standpoint of dynamics. Comparatively unsymmetric galaxies, characterised by a complex phase space that admits a coexistence of significant measures of both regular and chaotic orbits, can be surprisingly susceptible to low amplitude perturbations of the form that act in the real world and, as such, might be expected to exhibit systematic evolutionary effects over time scales $< t_D$. What precisely these evolutionary effects could be is not yet completely clear. However, there is good reason to believe that ongoing observational programs, such as the Sloan Digital Sky Survey, will provide at least partial answers to this basic question.

Acknowledgements

I am pleased to acknowledge useful collaborations with Robert Abernathy, Brendan Bradley, Barbara Eckstein, Salman Habib, Ilya Pogorelov, Ioannis Sideris, Christos Siopis, and, especially, Elaine Mahon, who first stimulated me to think about the issues discussed in this paper. This research was supported in part by NSF AST-0070809 and by the Institute for Geophysics and Planetary Physics at Los Alamos National Laboratory.

References

Batt, J.: 1987, *Transport Theory and Statistical Physics* **16**, 763.
Bohn, C. L., Sideris, I. V., Kandrup, H. E. and Kishek, R. A.: 2002, In *Proceedings of DESY 2002*, in press.
Chandrasekhar, S.: 1943, *Rev. Mod. Phys.* **15**, 1.
Contopoulos, G.: 1971, *Astron. J.* **76**, 147.
Copin, Y., Zhao, H. S. and de Zeeuw, P. T.: 2000, *Mon. Not. R. astr. Soc.* **381** 781.
Dehnen, W.: 1993, *Mon. Not. R. astr. Soc.* **265**, 250.
Evans, N. W.: 1994, *Mon. Not. R. astr. Soc.* **267**, 333.
Goodman, J., Heggie, D. and Hut, P.: 1993, *Astrophys. J.* **415**, 715.
Habib, S. and Ryne, R.: 1995, *Phys. Rev. Lett.* **74**, 70.

Habib, S., Kandrup, H. E., and Mahon, M. E.: 1997, *Astrophys. J.* **480**, 155.
Hunter, C., Terzic, B., Burns, A. M., Porchia, D. and Zink, C.: 1998, *Ann. N. Y. Acad. Sci.* **867**, 61.
Ipser, J. R. and Managan, R. A.: 1981; *Astrophys. J.* **250**, 362.
Kandrup, H. E.: 1998a, *Mon. Not. R. astr. Soc.* **299**, 1139.
Kandrup, H. E.: 1998b, *Mon. Not. R. astr. Soc.* **301**, 960.
Kandrup, H. E.: 1998c, *Astrophys. J.* **500**, 120.
Kandrup, H. E.: 2001, *Mon. Not. R. astr. Soc.* **323**, 681.
Kandrup, H. E., Abernathy, R. A. and Bradley, B. O.: 1995, *Phys. Rev.* **E51**, 5287.
Kandrup, H. E., and Mahon, M. E.: 1994, *Phys. Rev.* **E49**, 3735.
Kandrup, H. E., Pogorelov, I. V. and Sideris, I. V.: 2000, *Mon. Not. R. astr. Soc.* **311**, 719.
Kandrup, H. E. and Sideris, I. V.: 2002, *Celestial Mechanics* **82**, 61.
Kandrup, H. E. and Sideris, I. V.: 2001, *Phys. Rev.* **E64**, 056209.
Kandrup, H. E. and Siopis, C.: 2002, *Mon. Not. R. Astr. Soc.*, submitted.
Kishek, R. A., Bohn, C. L., Haber, I., O'Shea, P. G., Reiser, M. and Kandrup, H.: 2001, in *2001 IEEE Particle Accelerator Conference*, Evanston, IEEE Press, 151.
Kormendy, J. and Bender, R.: 1996, *Astrophys. J. Lett.* **464**, 119.
Lauer, T. R., Ajhar, E. A., Byun, Y.-I., Dressler, A., Faber, S. M., Grillmair, C., Kormendy, J., Richstone, D. and Tremaine, S.: 1995, *Astron. J.* **110**, 2622.
Lichtenberg, A. J. and Lieberman, M. A.: 1992, *Regular and Chaotic Orbits*, Springer, Berlin.
Lieberman, M. A. and Lichtenberg, A. J.: 1970, *Phys. Rev.* **A5**, 1852.
Louis, P. D. and Gerhard, O. E.: 1988, *Mon. Not. R. astr. Soc.* **233**, 337.
Mahon, M. E., Abernathy, R. A., Bradley, B. O. and Kandrup, H. E.: 1995, *Mon. Not. R. Astr. Soc.* **275**, 443.
Merritt, D. and Fridman, T.: 1996, *Astrophys. J.* **460**, 136.
Merritt, D. and Valluri, M.: 1996, *Astrophys. J.* **471**, 82.
Merritt, D. and Valluri, M.: 1999, *Astron. J.* **118**, 1177.
Miller, R. H. and Smith, B. F.: 1994, *Celestial Mechanics* **59**, 161.
Mineau, P., Feix, M. R. and Rouet, J. L.: *Astron. Astrophys.* **228**, 344.
Patsis, P. A., Athanassoula, E. and Quillen, A. C.: 1997, *Astrophys. J.* **483**, 731.
Perez, J. and Aly, J.-J.: 1996, *Mon. Not. R. Astr. Soc.* **280**, 689.
Pfaffelmoser, K.: 1992, *J. Diff. Eqns.* **95**, 281.
Pogorelov, I. V.: 2001, University of Florida Ph. D. thesis.
Pogorelov, I. V. and Kandrup, H. E.: 1999, *Phys. Rev.* **E60**, 1567.
Schaeffer, J.: 1991, *Commun. Part. Diff. Eqns.* **16**, 1313.
Schwarzschild, M.: 1979, *Astrophys. J.* **232**, 236.
Sideris, I. V. and Kandrup, H. E., 2001: *Phys. Rev.* **E65**, 066203.
Sinai, Ya.: 2000, private communication.
Siopis, C.: 2001, private communication.
Siopis, C. and Kandrup, H. E.: 2000, *Mon. Not. R. Astr. Soc.* **319**, 43.
Spitzer, L. and Schwarzschild, M.: 1951, *Astrophys. J.* **114**, 385.
Tennyson, J. J.: 1979, in Month, M. and Herrara, J. (eds.), *Nonlinear Dynamics and the Beam-Beam Interaction*, AIP Conf. Proc. **57**, p. 158
Tremblay, B. and Merritt, D.: 1995, *Astron. J.* **110**, 1039.
Udry, S. and Pfenniger, D.: 1988, *Astron. Astrophys.* **198**, 135.
Vandervoort, P. O.: 1980, *Astrophys. J.* **240**, 478.
Vandervoort, P. O.: 1984, *Astrophys. J.* **287**, 475.
van Kampen, N. G.: 1981, *Stochastic Processes in Chemistry and Physics*, Amsterdam, North Holland.
Vesperini, E. and Weinberg, M. D.: 2000, *Astrophys. J.* **534**, 598.

HOW IMPORTANT ARE THE EQUATIONS OF MOTION IN A CHAOTIC SYSTEM?

R. H. MILLER

Astronomy Department, University of Chicago, Chicago USA (E-mail: rhm@oddjob.uchicago.edu)

Abstract. We inquire whether some important features might be lost in studying the gravitational n−body problem because chaos makes it impossible to follow exact phase trajectories. We test this by comparing deliberately crude 'integrations' with published results from more precise methods. A general tendency found with high-quality integrations is that a central singularity forms at finite time. The time until that singularity forms is fairly reproducible for systems started from a more or less standard form. We find the same tendency with crude methods, although the time to reach that singularity from the usual initial conditions is less accurately reproduced with cruder methods.

1. Introduction and Background

The gravitational n−body problem is chaotic, as has been known since 1964 (Miller, 1964). By now, it is generally conceded that the e−folding rate for trajectory separation is several per crossing time (Goodman *et al.*, 1993; Kandrup *et al.*, 1994). A crossing time is the dynamical time scale for a stellar system, or a radian of orbital motion, etc. The e−folding rate is essentially the largest Lyapunov characteristic exponent. The gravitational n−body problem is extremely sensitive to initial conditions.

Picture the n−body system as represented by a system point in its $6n$−dimensional phase space, which is a differentiable manifold. The system has 10 first integrals of motion: three for the centroid position, three for the centroid momentum, three for angular momentum, and one for the total energy. The value of each of these integrals remains constant as the motion proceeds. There is a $(6n - 10)$−dimensional sub-manifold embedded in the $6n$−dimensional phase space on which these integrals all have their 'proper' value. In earlier days, I referred to that submanifold as 'the integral hypersurface.'

While the integrals themselves are not independent in the sense of involution (their Poisson brackets are not all zero), their gradients are linearly independent. A patch tangent to the submanifold is orthogonal to the gradients of all 10 integrals, evaluated for the point in question.

In real calculations, the system point drifts off the integral hypersurface. Some years back, I reported a strategy to move the phase point back onto that hypersurface from time to time as an n−body problem runs (Miller, 1971b).

The basic idea is to move a phase point along a normal toward the integral hypersurface, where that normal is defined by the gradients of the first 10 integrals. For the refinements, the closest point on the integral hypersurface was found by using generalized inverses for the (non-square) $(6n \times 10)$ matrix of the gradients. This process is coupled to an ordinary integrator, and it is carried out after each integration step. The process requires iteration to allow for local curvature of the integral hypersurface, so I referred to it as 'Partial Iterative Refinements.'

My original paper appeared along with a companion paper in which the method was used to explore the way in which the computed phase point swims around over the integral hypersurface (Miller, 1971a). This was done in the usual sense of a 'difference calculation' in which two calculations are both run forward in time, starting from phase points initially very close together and watching the difference.

The difference vector stays close to the integral hypersurface, but it swims around within that surface. The direction–cosine of the difference vector onto the the trajectory swings from positive to negative. Most of the time the cosine stays close to ± 1 (with due allowance for the large dimensionality of the space), but the difference vector swings from parallel to antiparallel. It spends about as much time antiparallel as parallel. This even happens when the initial difference vector lies along the trajectory; such a system can develop difference vectors that swing from parallel to antiparallel only through numerical errors. This happens in computed systems.

The e—folding rate for the length of the difference vector is not significantly affected by the refinement procedure. These refinements don't improve an n—body calculation, a conclusion that might be expected given the large number of dimensions of the submanifold on which the integrals have their proper value.

Paul Nacozy independently reported the same idea, except that he didn't iterate the process to allow for curvatures in the integral hypersurface. The two of us disagreed on whether the process actually helps with n—body calculations. Nacozy's paper appeared both in the *Proceedings* of the 1970 Cambridge n—body conference (Nacozy, 1972) and in *Astronomy & Space Science* (Nacozy, 1971).

I recognized at the time that one could fake an n—body calculation by doing something like a random walk in the phase space, requiring that each step lie in the tangent space so the phase point remains near the integral hypersurface. From time to time, the phase point can be shoved back onto the surface by the iteration process if necessary. This conference finally nudged me to try it out, and what follows is a report on those attempts.

Attention is directed at simple calculations of early stages of star cluster development in this note. Special cases might be imagined that, for example, admit additional integrals beyond the original ten. But the methods used here require that all gradients of this expanded set of integrals be linearly independent, a property that is not at all evident *a priori*.

The exercise described here piqued my curiosity, and it was fun to work on. This method would never be used as a substitute for a real n-body calculation. It is even more expensive to run.

2. The Question

Such a faked system manifestly pays little attention to the equations of motion, so the question may be posed: just how important are the equations of motion, once the first integrals are constrained to their initial values?

And, given that the system is chaotic, how important is that Chaos? What effect does it have on practical integrations? In fact, we'll not answer this question. Instead, we use that chaotic property to explore the other questions.

Of course, the first integrals are constant as a consequence of the equations of motion. That changes the question slightly to inquire what additional properties of the motion are forced upon a system by the equations of motion. The systems may simply be seeking regions of the phase space that are consistent with the integrals and are, in some sense, most probable. Such a process is usually described as 'maximizing available phase volume.' These systems are of such complexity that resonance structures, etc., are likely to be so finely divided that one may feel fairly safe in ignoring them.

Another point of interest in any physical problem is how long it takes a system to pass from its starting point to those interesting 'most probable' regions. Initial conditions (the starting point) are pretty arbitrary. Statistical mechanics doesn't provide much help in this area. It's the old problem of 'the approach to equilibrium' that has plagued statistical mechanicians for a long time.

3. Some Details

The strategy is as follows. Given a phase point, that phase point is to be given a small step in some 'random' direction within the tangent space. The move yields a new phase point, which lies near the integral hypersurface. The process is repeated over and over again. This is a bit like an integration, but each 'integration step' is simply a step in a random walk. The system is constrained to remain near the integral hypersurface by making each step orthogonal to all the gradients to that hypersurface.

A strict random walk is not a good method to seek regions of greater phase volume. Random walks build a cloud of points that remain near the point where the walk started. They seldom wander far away from that initial point. An inordinate number of steps would be required for the phase point to go very far. The walk needs some help to keep it going more or less in the same direction.

There are also problems deciding how long a step can safely be taken. The tangent space departs too far from the integral hypersurface if steps are too large. The surface becomes strongly curved if several particles get close to each other.

Something that mixes bits and pieces of an integration with a random walk will seek interesting regions of the phase space, because phase trajectories under the equations of motion can explore distant regions of the phase space. One might imagine doing a bit of an integration, then taking a small random walk step (confined to the tangent space), then another bit of integration, and so on. The problem of how large a step of the random walk to use again plagues this approach.

A quick and dirty scheme to do that is simply to use an integration method deliberately chosen to be crude. For the work reported here, that was the procedure. A leapfrog integrator was used, in a form which allows variable time steps. The crude integration is coupled with iterative refinements, since integration errors allow the system to drift away from the integral hypersurface. Refinements are made whenever the integrals drift too far from their initial values. The leapfrog is symplectic, but that's probably not important in the present circumstances. The refinement procedure probably cannot be made symplectic. Other, more elaborate, schemes for moving a system point around in the phase space are easy to imagine, but this seemed to be a good place to start.

4. Comparisons With Real n-body Calculations

In the best of worlds, we might look for some manner in which results from a calculation like this differ from a real n−body calculation.

A study often cited to make the case that n−body calculations are valid is that of Aarseth, Hénon, and Wielen (1974), in which results from collisional n−body calculations (Aarseth, Wielen) were systematically compared against specialized Monte Carlo calculations by Hénon and by Spitzer and against hydrodynamic calculations by Larson. Hénon (1973) gave a reasonably full account of the Monte Carlo method, and Spitzer described his approach, and its reconciliation with Hénon's, in his book on Globular Clusters (Spitzer, 1987).

The Monte Carlo method enforces the integrals of motion, but otherwise it pays no attention to any other consequences of the equations of motion. In enforcing the integrals, particular emphasis is placed on energy and angular momentum. The random element in these Monte Carlo calculations is applied at any time step through transitions that mimic the pairwise encounters of 'superparticles.'

Even so, it is of some interest to revisit this problem, to see if something more can be learned.

A key to the present study is to use the same model and same analysis methods as were used by Aarseth, Hénon, and Wielen (1974). Their calculations started with a Plummer model, whose length scale and total energy were set by Aarseth's convention wherein the gravitational constant and the total mass are set to unity and the total energy $\mathcal{E} = -\frac{1}{4}$. Times for their plots are measured in units of relaxation times for the initial cluster, and length units are given by an initial cluster radius.

Their systems tended toward a state of high central concentration, a precursor to the 'gravothermal oscillations' now so prominent in globular cluster studies. The time to reach that state is of particular interest. In 1974, when Aarseth, Hénon, and Wielen wrote their paper, integrations could not be extended past that central 'singularity,' which was reached as the central concentration got greater and greater. That is no longer a limitation today, but the methods employed do not bear on the question raised in this paper.

5. Results

Several exploratory experiments are summarized here.

First, my models showed the same general trends as those displayed by Aarseth, Hénon, and Wielen (1974). Lagrangian radii at 10%, 50%, and 90% of the total mass trended in the same directions. The principal difference is that the systems in some experiments got in trouble with small values for the 10% Lagrangian radius at earlier times than shown by Aarseth *et al.*, while others did it later. Experiments that used shorter time steps, thus a better integration, show the 10% Lagrangian radius collapsing at about the same time as the Aarseth *et al.* systems, while experiments that use cruder methods (here obtained by using longer time steps in the integrator) depart more strongly. The conclusion here is that the times required to reach a central singularity are indicated more reliably by careful integrations than with the fake methods. Increasingly sloppy methods get worse and worse.

Second, these crude methods couldn't handle close binaries very well. The principal difficulty seems to be that the topology of the integral hypersurface gets very tortured when local particle densities get large, especially if a reasonably tight binary is about to be troubled by a passing star. This might be remedied with more careful treatments of the iteration process. The symptoms were that the refinement process didn't converge within some allowed tolerance within 10 tries. In a few cases, it converged for a while, then it suddenly started to show *larger* total errors in the integrals.

6. Discussion

The equations of motion seem to do little beyond controlling the rate at which the system reaches a state with strong central concentration, once conservation of the first integrals is ensured. No new features are evident in the results of careful n-body integrations that are not seen with cruder integrations. Dramatic differences would be required to make a convincing case that crude integrations miss interesting features in the problem, but no substantial differences were found. The agreement seen in the data presented by Aarseth, Hénon, and Wielen (1974), using various methods in addition to n−body integrations, reinforces this conclusion.

Acknowledgements

Most calculations for this program were carried out at the NASA-Ames Research Center's NAS Systems Division under an HPCC grant of computing resources, which are gratefully acknowledged. Support from Cooperative Agreement NCC2-1103 with the NASA-Ames Research Center is also appreciated.

References

Aarseth, S. J., Hénon, M. and Wielen, R.: 1974, 'A Comparison of Numerical Methods for the Study of Star Cluster Dynamics'. *A&A* **37**, 183.

Goodman, J., Heggie, D. C. and Hut, P.: 1993, 'On the Exponential Instability of n-body systems'. *Astrophys. J.* **415**, 715, Fiche 196–F11.

Hénon, M.: 1973, 'Collisional Dynamics in Spherical Stellar Systems', in: Martinet, L. and Mayor, L. (eds.), *'Dynamical Structure and Evolution of Stellar Systems'*, *Third Advanced Course, Swiss Society of Astronomy and Astrophysics*. Sauverny, pp. 183–260, Geneva Observatory.

Kandrup, H. E., Mahon, M. E. and Smith, H.: 1994, 'On the Sensitivity of the n-Body Problem Toward Small Changes in Initial Conditions', *Astrophys. J.* **428**, 458, Fiche 135–G12.

Miller, R. H.: 1964, 'Irreversibility in Small Stellar Dynamical Systems', *Astrophys. J.* **140**, 250.

Miller, R. H.: 1971a, 'Experimental Studies of the Numerical Stability of the Gravitational n-Body Problem', *J. Comp. Phys.* **8**, 449.

Miller, R. H.: 1971b, 'Partial Iterative Refinements', *J. Comp. Phys.* **8**, 464.

Nacozy, P. E.: 1971, 'The Use of Integrals in Numerical Integrations of the N-Body Problem', *Ap&SS* **14**, 40.

Nacozy, P. E.: 1972, 'The use of Integrals in Numerical Integration of the N-Body Problem', in: Lecar, M. (ed.), *'Gravitational n-body Problem'*, *Proceedings of IAU Colloquium No. 10*. Dordrecht, pp. 153–164, Reidel.

Spitzer, L.: 1987, Dynamical Evolution of Globular Clusters. Princeton, NJ: Princeton University Press.

RESONANTLY EXCITED NON-LINEAR DENSITY WAVES IN DISK SYSTEMS

CHI YUAN

Institute of Astronomy & Astrophysics, Academia Sinica, Taipei, Taiwan, ROC

Abstract. Most of the disk systems are characterized by spiral structures. A good portion of these spiral structures can be identified as waves resonantly excited by a perturber in or pertaining to the same system. For planetary rings, this is an exterior satellite; for galactic disks, a rotating bar; for proto-stellar disks (yet to be confirmed), this would be a proto-planet. These waves, not just responsible for the present morphology of the disks, also play a dominating role on evolution of the disks. Resonance excitation is a extremely effective mechanism. A relatively weak perturbation can result in a highly non-linear responses in the disk. Therefore, non-linear theory is a necessity here. We will examine the non-linear theory of resonance excitation and discuss the applications of the theory to Saturn's rings and disk galaxies in this paper.

1. Introduction

Disks are perhaps the second most common form in the Universe. They range from 10^{10} cm for planetary rings, to about 10^{15} cm for proto-stellar disks and 10^{13-16} cm for various kinds of accretion disks, and to 10^{22} cm for galactic disks. Their sizes span over 10^{12} in magnitude. Yet, amazingly, they are marked by the same spiral structure in their appearance. Now we know, as first pointed out by Lindblad (1963), the spiral structure is of quasi-stationary nature. This concept was later adopted by Lin and Shu (1964, 1966) and these spirals are now generally referred as density waves. They are common in rapidly rotating systems in nature. The density wave theory has since been developed into a major and mature field in astrophysics.

The development of the theory has taken two different directions. One is to emphasize the self-excited mechanism, to treat the spirals as the modes of a dynamical disk system. This approach has been adopted by C.C. Lin and his co-workers (Bertin *et al.*, 1989a, 1989b). The theory and its applications are best summarized in recent books by Bertin and Lin (1997) and Bertin (2000). The other is to invoke a resonance mechanism to interpret the spirals as waves excited by a periodic perturber. This approach is attributed to P. Goldreich and S. Tremaine in their treatment of Saturn's rings (1978a, 1978b, 1979). In that case, the perturber is a satellite exterior to the rings. The same idea was developed for the spirals in barred galaxies (Goldreich and Tremaine, 1980), in which a bar is the perturber. A non-linear theory based asymptotic approach, but solved by numerical hydrodynamics was developed and applied to the '3 kpc Arm' problem for the Milky Way (Yuan

Space Science Reviews **102**: 121–138, 2002.

1984). Similar studies have extended application to the study of proto-stellar disks (Lin and Papalozou, 1986; Yuan and Cassen, 1992). This two approaches which address different problems are complementary to each other. Take disk galaxies for instance. For an isolating galaxy, the spiral structure may be explained by the modal approach. For barred galaxies or galaxies with a central (or nuclear) bar, the spirals are those resonantly excited by the bar.

The purpose of the present article is to review the development of the second approach, the resonance excitation mechanism and its applications. We refer the recent development of the first (modal) approach to the excellent review books by Bertin and Lin as quoted above. Neither shall we be involved in discussing the amplification mechanism of the waves, a fundamental issue in the density wave theory, except mentioning that a recent work of Shu *et al.* (2000) seems to indicate unification of rivalry processes, the waser (Mark, 1976) and the swing amplifier (Toomre, 1981) is plausible. This supports the view maintained for years by C.C. Lin that these two processes co-exist and that the over-reflection in waser at corotation is equivalent to the swing amplification for evolving waves there.

The spiral waves are excited at all three Lindblad resonances, the outer, the outer inner and the inner inner. For convenience, they are respectively referred as OLR, OILR and IILR. So far, the observations of waves associated with IILR have not been firmly established. For OLR and OILR, however, cases are numerous. The resonance excitation is an effective process. A forcing by the perturber with a strength of 2% of the mean field at the resonance can produce highly non-linear density waves. These waves propagate mainly by means of self-gravitation of the disk as in planetary rings, or as acoustic waves as in proto-stellar disks, or as a combination of both as in galactic disks. These non-linear waves are eventually attenuated by viscosity. Thus to formulate the problem, we must include effects of self-gravitation, pressure gradients, and viscosity. Thus it would require the full gasdynamic equations coupled with the Poisson equation. Orbital calculations of individual particles or N-body calculations can provide in important physical in-sights to the problem, but cannot give the full picture, at least not yet. We will refer the recent development of these areas to the article by G. Contoupolous, P. Grosbol and R. Miller in this volume.

The gasdynamic equations are well known to be difficult to handle. There are two different approaches: one to solve the equations by numerical simulations and the other, by using asymptotic analysis. There are advantages and disadvantages of both approaches. The numerical simulation is perhaps the only approach which can tell the evolution of the disks, but it does not shed light on the physical processes involved with the resonance excitation. The asymptotic analysis approach, on the other hand, can reveal the physical processes, but is incapable of solving the evol-ution problem. In this article, we choose to use the asymptotic analysis, focusing on the physical processes involved in the resonance excitation. Another reason we use this approach is that we can solve problems of planetary rings, galactic disks, and proto-stellar disks with the same formulation. For the numerical simulations, there are several conference proceedings dedicated to the subject with references within, which we quote: Nobel Symposium 98: Barred Galaxies and Circumnuclear Activity (Sandqvist & Lindblad, 1996), IAU Colloquium 157: Barred Galaxies (Buta et al., 1996), NAP98: Numerical Astrophysics (Miyama et al., 1999), and 15th IAP Meeting Dynamics of Galaxies (Combes et al., 2000).

We first describe the resonance excitation mechanism in section 2, and then we outline the Lagrangian formulation we adopt in section 3. In section 4, we review the problem of Saturn's rings. In section 5, we examine the bar-driven spiral density waves in disk galaxies, especially for the gas-dust disks in the central regions. We give a brief report on waves in proto-stellar disks in section 6. Some concluding remarks are to be found in section 7.

2. Resonance Recitation Mechanism

The issue that spiral density waves can exist is intimately related to the disk instability. In the self-excited case, one looks for wave solutions through a boundary value problem, in which eigenmodes with eigenvalues $\omega = m\Omega_p - i\gamma$ are obtained. Spiral waves with m arms would propagate circumferentially with pattern speed Ω_p and get amplified at a rate $e^{\gamma t}$. For resonantly excited waves, Ω_p is provided by the perturber and amplitude of the excited waves also depends solely on the perturber. The growth rate is no more an issue. For bar-driving cases, Ω_p is the angular speed of the bar. For planetary rings or proto-stellar disks, it is given by,

$$Re(\omega) = m\Omega_p = m\Omega_s \pm l\kappa_s \pm p\mu_s,$$

where Ω_s is the angular speed of a satellite or a proto-planet, κ_s, its epicyclic frequency. Furthermore since often the perturber does not lie on the same plane of the ring or the proto-stellar disk, it would also undergo vertical oscillations. μ_s represents the vertical frequency. The quantities m, l, p are positive integers and the subscript s denotes the secondary, which is either satellite or proto-planet. When a particle at distance r from the center of the disk satisfies the following relations:

$$Re(\omega) - m\Omega(r) = \pm\kappa(r), \quad Re(\omega) - m\Omega(r) = \pm\mu(r),$$

$r = r_L$ is the location of the Lindblad resonance for the first relation and $r = r_V$ is the location of bending resonance for the second relation, where $\Omega(r)$ is the rotation of the system and $\kappa(r)$ is the local epicyclic frequency. The plus (minus) sign on the right side of the above relations indicates the resonance is outside (inside) the co-rotation, hence outer (inner) Lindblad or bending resonance, where the co-rotation (CR) is defined by distance r such that $\Omega(r) = \Omega_p$. The meaning of the above equations is clear that the particle at $r = r_L$ ($r = r_V$), oscillates radially (vertically) at the same frequency it sees the perturber. This gives rise to resonance excitation.

The physical mechanism can be best understood from a linear analysis. The model equation which governs the motion of the particle is none other than the familiar Klein-Gordon equation (modified by the self-gravitation term) in the WKBJ approximation. By denoting the radial displacement $z(r, t)e^{-im\theta}$, the equation can be written as (Yuan and Cheng, 1989):

$$\frac{\partial^2 z}{\partial t^2} - a^2 \frac{\partial^2 z}{\partial r^2} + 2\pi G\sigma_o i \frac{\partial z}{\partial r} + \kappa^2 z = \frac{\Psi_1}{r} e^{i\omega' t}. \tag{1}$$

In the above equation, σ_o is the ambient surface density of the disk, a is the sound speed of the gas medium, Ψ_1/r is amplitude of the effective forcing, and $\omega' = \omega - m\Omega$, the frequency that the particle sees the perturber. Without the self-gravitating term, the equation is regular inhomogeneous Klein-Gordon equation. The resonance occurs at r where $\omega' = \kappa$. The excited particles would organize into density waves, propagating with acoustic speed a. Once the self-gravitation effect is included, the nature of the wave changes. This becomes clear once we study the dispersion relation.

In the linear theory, the azimuthal displacement ϕ in the frame rotating at Ω_p can be approximated by $\phi = \phi_0 + (\Omega - \Omega_p)t$ with $\phi = \theta - \Omega_p t$. Thus, we may write:

$$\frac{\partial^2 z}{\partial t^2} = -m^2(\Omega - \Omega_p)^2 z = -(\omega - m\Omega)^2 z.$$

Then, we obtain the governing equation[1] for the linear density waves:

$$-a^2 \frac{d^2 R}{dr^2} + 2\pi G\sigma_o i \frac{dR}{dr} + [\kappa^2 - (\omega - m\Omega)^2]R = \frac{\Psi_1}{r}, \tag{2}$$

where $z = R(r)e^{i(\omega - m\Omega)t}$. In the WKBJ approximation, we assume

$$R(r) = A(r)e^{i \int k(r)dr},$$

where $A(r)$ is the slowly varying amplitude and $\int k(r)dr$ is the rapidly varying phase with $k(r)$ the radial wavenumber of the spiral waves, and $|kr| \gg 1$, i.e., in the limit of tightly wound spirals. Using the formula for R above, we obtain the well known Lin-Shu dispersion relation:

$$a^2 k^2 - 2\pi G\sigma_0 |k| + [\kappa^2 - (\omega - m\Omega)^2] = 0. \tag{3}$$

This simple relation, amazingly, contains essential information about wave propagation for all disk systems[2]. We will not repeat the well known results of this relation here. However, we do want to point out that, except for the IILR which so far has not be firmly identified in observation, resonance excitation uniquely picks the trailing waves. We may summarize the scenario: *Long trailing waves are excited at the Lindblad resonances, propagate toward the co-rotation, and, before reaching it, are refracted (reflected) as short trailing waves at Q-barrier, and propagate toward*

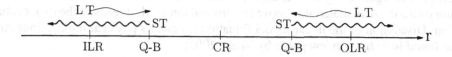

Figure 1. Scenario for resonantly excited waves. Long trailing waves (LT) are excited at OLR and ILR, propagate toward co-rotation (CR), and reflected or refracted as short trailing waves at the Q-barrier (Q-B), propagating toward and beyond the Lindblad resonance where they are generated

and pass the original Lindblad resonance and beyond. The Q-barrier can be understood as follows: The transformation of the long trailing waves to short trailing waves occurs in the radius, where the radial group velocity of the wave tends to zero, $\partial\omega/\partial k = 0$. The latter equality asserts, as it follows from equation (3), at the Q-barrier – the location separating the wave region and the forbidden region – where $k = k_0 = \pi G\sigma_0/a^2$. Substituting the expression for k_0 in Equation (3), we obtain

$$(\pi G\sigma_0)^2 - a^2\left[\kappa^2 - (\omega - m\Omega)^2\right] = 0.$$

This scenario is depicted in Figure 1.

The direction of the wave propagation in Figure 1 is determined by (a) the conservation of the angular momentum flux of the waves (Toomre, 1969, for stellar disks; Goldreich and Tremaine, 1979 for gas disks),

$$\frac{d}{dr}[r\Phi^2(1 - \frac{a^2|k|}{\pi G\sigma})] = 0, \tag{4}$$

with Φ, the amplitude of the spiral gravitational potential which is proportional to amplitude A of the displacement R, and (b) the group velocity,

$$c_g = -\frac{d\omega}{dk} = -sgn(k)[\frac{\pi G\sigma - |k|a^2}{m(\Omega - \Omega_p)}], \tag{5}$$

where we take $\omega = m\Omega_p$. With these relations, the waves's excess angular momentum density[3], H, to be referred later is

$$H = -\frac{\Omega - \Omega_p}{2\sigma}(\frac{m\Phi}{2\pi G})^2. \quad^1$$

Solving Equation (2) with proper boundary conditions would give us the total picture how linear waves are generated at r_L (r_V) and propagate away from it afterwards. It is consistent with the above scenario. Integrating $Im(R)$ will give us the torque that the perturber exerts on the disk. Conversely, the gravitational force of the spiral density waves in the disk exerts an equal but opposite torque back to the perturber. These waves, attenuated by viscosity, would deposit the excessive (negative or positive) angular momentum in the wavy region on the disk, and cause the disk material to move inward and outward. It leads to the disk evolution. Discussion of the linear model Equation (2) and its physical implications can be found in a short review paper by Yuan (2001).

3. The Non-linear Equation for Resonantly Excited Waves

In this section, we shall introduce a non-linear singular integro-differential equation, which computes the response of the disk to a periodic forcing and gives the results we are looking for. Given the limited space here, we cannot present the derivation, but simply state the results and their physical implications. The full derivation can be found in Shu *et al.* (1985) and Yuan & Cheng (1989). The disk is assumed to be thin, self-gravitating, and viscous, and the fluids in it follow either an isothermal or a polytropic law. And we are interested in steady state solutions. The problem can be readily formulated in the familiar Eulerian approach. However, in order to capture essential features of the non-linearity without being entangled in the full-scale numerical simulations, it is more convenient to formulate the problem by the Lagrangian approach. There are several distinctive advantages in the asymptotic limit $|kr| \gg 1$, by using the Lagrangian formulation: (1) Simpler equation of continuity, (2) Simpler Poisson equation, (3) No convection terms in the equation of motion, and (4) Streamlines as direct results. The disadvantages are more complicate stress tensors, for pressure and viscosity.

Before we write down our results, we first introduce the variables in the Lagrangian formulation. We specify the radial and azimuthal displacements as:

$$r = r_0 + r_1(r_0, \phi_0),$$

$$\phi = \phi_0 + [\Omega(r_0) - \Omega_p]t + \phi_1(r_0, \phi_0).$$

The coordinates (r_0, ϕ_0) are the independent variables and (r_1, ϕ_1) are the dependent variables. However, it is more convenient, for multiple-arm spirals to have,

$$\psi_0 = m\{\phi_0 + [\Omega(r_0) - \Omega_p]t\},$$

$$\psi = m\phi = \psi_0 + \psi_1(r_0, \psi_0).$$

With these variables, we can express, in the WKBJ approximation, the equation of motion as (Shu *et al.*, 1985):

$$\sigma(r, \psi) = \frac{\sigma_0(r_0)}{1 + \dfrac{\partial r_1}{\partial r_0}}, \tag{6}$$

and the Poisson equation (Shu, 1984)

$$\frac{\partial \mathcal{V}_d}{\partial r} = 2G \int_0^\infty \frac{\sigma(r', \psi_0)}{r - r'} dr', \tag{7}$$

where $\mathcal{V}_d$ is the potential of the self-gravitating disk. The perturbation potential $\mathcal{V}_1$, in terms of ψ, can be written as $\mathcal{V}_1 = V_1\cos\psi$. The equations of motion are more conveniently written for r_1 and J_1, where J_1 is the perturbed angular momentum, such that $J = J_0 + J_1$:

$$(\omega - m\Omega)^2 \frac{\partial^2 r_1}{\partial \psi_0^2} + \kappa^2(r_0)r_1 = 2\Omega(r_0)\frac{J_1}{r_0} - \frac{dV_1}{dr_0}\cos\psi_0 - \frac{\partial\mathcal{V}_d}{\partial r} - \frac{1}{\sigma_0}\frac{\partial P_{rr}}{\partial r_0}, \quad (8)$$

$$-(\omega - m\Omega)\frac{\partial J_1}{\partial \psi_0} = mV_1\sin\psi_0 - \frac{r_0}{\sigma_0}\frac{\partial P_{r\phi}}{\partial r_0}, \quad (9)$$

in which we replace the time derivative by ψ_0 derivative, using the relation between ψ_0 and t above, and $(P_{rr}, P_{r\phi})$ are the stress tensor for both pressure and viscosity. Equations (6)–(9) are the basic equations for the problem.

Now we need to take two steps: (1) to expand r_1 in series of a small parameter and solve the problem iteratively and (2) to impose the condition of no secular terms in each iteration (actually only twice) from integrating the equation over ψ_0. This will give us the non-linear equation we need for the resonance excitation. The small parameters, respectively for planetary rings, and for proto-stellar disks or galactic disks are

$$\epsilon = \frac{2\pi G\sigma_o}{|\mathcal{D}|r_L}, \quad \delta = \left|\frac{a^2}{r_L^2 \mathcal{D}}\right|,$$

in which

$$\mathcal{D} = \left(r\frac{dD}{dr}\right)_{r=r_L}, \quad D = \kappa^2 - (\omega - m\Omega)^2.$$

The quantity ϵ represents the ratio of the ring mass to the planet mass, and δ is roughly the square of the ratio of sound speed to the rotation speed. The resulting equation for dimensional displacement R such that $r_1 = R(r_0)\exp^{i(\omega t - \psi_0)}$, is

$$\frac{d}{dr_0}\left\{a^2\left[I(q_0^2) + ib(q_0^2)\right]\frac{dR}{dr_0}\right\} - 2G\sigma_0\int_{-\infty}^{\infty} I(q^2)\frac{R(r_0') - R(r_0)}{(r_0' - r_0)^2}dr_0'$$

$$+\left[\kappa^2 - (\omega - m\Omega)^2\right]R(r_0) = \frac{\Psi_1}{r_0}. \quad (10)$$

The nonlinearity comes in through the non-linear functions I for pressure and b for viscosity. They are functions of the streamline-crossing parameters q_0 and q, which, always less than 1, are defined as

$$q_0 = \left|\frac{dR}{dr_0}\right|, \quad q = \left|\frac{R(r_0') - R(r_0)}{r_0' - r_0}\right|.$$

Functions I and b are defined as:

$$I(q^2) = \frac{2}{q^2}[(1 - q^2)^{-\frac{1}{2}} - 1],$$

$$b(q_0^2) = \frac{\kappa}{\sigma_0(r_L)a^2}\left\{(\frac{4}{3}\mu + \mu_1)[I_0(q_0^2) - I_2(q_0^2)] + \mu I(q_0^2)\right\},$$

with

$$I_n(q_0^2) = \frac{1}{q_0^n(1 - q_0^2)^{\frac{1}{2}}}\left[(1 - q_0^2)^{\frac{1}{2}} - 1\right]^n, \quad n = 0, 1, 2.$$

The quantities, μ and μ_1, are, respectively, the shear and bulk coefficients of viscosity. Without going details in solving this equation, we make a few remarks below.

– Equation (10) will become the linear equation (2) in the limit $q_0 \to 0$ and $q \to 0$.

– Although Equation (10) correctly describes the non-linear wave motions in response to a periodic perturber, it is very difficult to solve because of its singular integral nature. Recognizing the contribution to the integral comes mainly from the neighborhood of the singularity, we can replace the singular integral by a differential operator, using a local approximation for the integral (Shu *et al.*, 1985; Yuan & Cheng 1991). The new equation satisfies the conservation of angular momentum flux carried by the waves and gives a good approximation to Equation (10) when the perturbation field is a few percent of the mean field at resonances, which turns out to be the case for all the problems under consideration. The resulting equation is:

$$\frac{d}{dr_0}\left\{[I(q_0^2) + ib(q_0^2)]\frac{dR}{dr_0}\right\} + i4G\sigma_0 L^{1/2}\frac{d}{dr_0}(L^{1/2}R)$$

$$+ \left[\kappa^2 - (\omega - m\Omega)^2\right]R(r_0) = \frac{\Psi_1}{r_0}, \qquad (11)$$

where

$$L(q_0^2) = \frac{1}{\pi}\int_0^\infty I(q^2)\frac{\sin 2\zeta}{\zeta}d\zeta.$$

– The forcing term appearing on the right hand side of equation (10) contains two parts:

$$\frac{\Psi_1}{r_0} = -\frac{dV_1}{dr_0} + \frac{2m\Omega(r_0)}{\omega - m\Omega(r_0)}\frac{V_1}{r_0}.$$

The first part is the gravitational force due to the perturber and the second part is the Coriolis force associated with the circular motion of the perturber. Usually the latter dominates. Since Coriolis force changes sign across the co-rotation, it is in the same direction of the gravitational force of the perturber at OLR, hence reinforcing it, but in the opposite direct at ILR, hence reversing the direction

of the gravitational pull. This suggests that relative to resonance excitation at
OLR, waves at ILR are weak and out of phase of the bar potential.

- The torque that the perturber exerts on the disk, according to this non-linear
theory is:

$$T = -m\psi_0(r_L)\Psi_1(r_L) \int_{-\infty}^{\infty} Im(R)\, dr_0. \tag{12}$$

This formula enables us to calculate the rate that the perturber transports its
angular momentum to the disk and gives us the estimate of the lifespan of the
disk. Since the sign of $\Psi_1(r_L)$ is negative for OLR and positive for ILR. Thus,
the angular momentum density of the spiral waves is positive for OLR, and
negative for ILR (both OILR and IILR). Accordingly, the disk material, gaining
angular momentum, will move out in the region covered by the spiral density
waves excited at OLR, and will move in in the region covered by the spiral
waves excited at OILR and IILR. The former is responsible for the expansion
of the 3-kpc arm in the Milky Way (Yuan, 1984), for which a rapidly rotating
bar was first assumed by Yuan in the paper quoted above and later observed
(e.g., Blitz and Spergel, 1990). The latter is responsible for the fueling the
AGN (Shlosman et al., 1990), if the spirals can reach all the way to the center.
Otherwise, the material will accumulate into a ring-like structure which may be
triggered into star burst activities (Yuan & Kuo, 1997, 1999).

4. Saturn's Rings

A notable success of the theory of resonance excitation is found in its application
to Saturn's rings. More than 50 wavetrains, both density waves and bending waves,
have been identified in Saturn's rings. Their features are in good agreement with the
non-linear theory of density waves (Shu et al., 1985b). Equally successful are the
bending waves (Shu et al., 1983). However, non-linear theory for bending waves
have not been developed. According to the theory, the spiral waves excited at the
ILR will propagate outward and be attenuated by the viscosity of the Saturn's
rings, and the negative excess angular momentum of the waves would be deposited
in the neighborhood outside the ILR, which is covered by the waves, and would
eventually remove disk material to form a gap. This mechanism certainly works
for a good number of gaps in Saturn's rings. Whether it can clear a wide gap such
as Cassini Division remains controversy (See the critism and a different approach
by Fridman & Gorkavyi, 1999).

For Saturn's rings, the dispersion velocity of the ring particles, which is equivalent to the sound speed in gas systems, is very small compared with the rotation
speed. It is of the order of 1 mm/s or less. In practice, we may ignore it[4]. On the
other hand, the viscous effect cannot be ignored. Furthermore the collision frequency of the ring particles is comparable to the rotation frequency. The viscosity

therefore cannot be determined as in ordinary fluids, as we did in formulating our problem. A crude formula was used by Goldreich and Tremaine (1978a), and more careful one was due to Borderies *et al.* (1984), which was simplified by Shu and Steward (1985). With these consideration, the governing equation becomes:

$$-2G\sigma_0 \int_{-\infty}^{\infty} I(q^2) \frac{R(r_0') - R(r_0)}{(r_0' - r_0)^2} dr_0' + \left[\kappa^2 - (\omega - m\Omega)^2\right] R(r_0)$$

$$= \frac{\Psi_1}{r_0} + \frac{dF}{dr_0}, \tag{13}$$

where F is a stress function of R. Again this equation will be replaced by a differential equation, similar to equation (11). The torque is determined by Equation (12). Shu *et al.* (1985b) solved Equation (13), using the same heurestic approach as in deriving Equation (11). A typical result for the surface density is shown in Figure 2. Qualitatively it agrees with the observations of the waves at Mimas 5:3 ILR, shown in Figure 3.

It might be noted that spiral waves in Saturn's rings are always long trailing waves. Short waves are in fact shorter than the size of Saturn's ring particles, hence cannot possibly exist. Long trailing waves are excited at the ILR and propagate outward, but they are attenuated by viscosity in a distance of a few hundred km and never reach the Q-barrier, which are tens of thousand km away.

5. Bar-driven Spiral Waves in Galaxies

Although obvious that spirals and bars are closely related in barred galaxies, understanding this relationship and the physical mechanism behind the scene came much later. First evidence came from the numerical simulation showing surprisingly, that spiral density waves could be generated by a rotating bar potential (Huntley, Sanders, & Roberts 1978). A systematic study of the mechanism, however, was not available until a year later (Goldreich and Tremaine, 1979, 1980). The mechanism is the same as in planetary rings. The difference is that in the case of disk galaxies, the waves are supported by both pressure and self-gravity of the disk, while in Saturn's ring, mainly by self-gravity. Acoustic waves (pressure) are basically short waves in nature and gravitational waves are long waves. The gravitational forcing by the perturber necessarily excites long waves at Lindblad resonances. Since only trailing can propagate away from the Lindblad resonance, it is necessarily long trailing waves. At the Q-barrier, the long trailing waves are transformed[5] into short trailing waves. Only short trailing waves can leave the Q-barrier and travel beyond the original Lindblad resonance. In the papers quoted above, Goldreich and Tremaine derive the linear torque formula and show how the angular momentum is transported between the disk and the bar via spiral density waves (confirming earlier results of Lynden-Bell and Kalnajs, 1972).

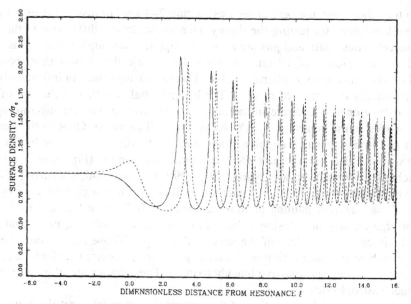

Figure 2. Density profiles calculated from equation (13) for waves excited at Mimas 5:3 ILR. Two profiles are radial cuts at two phase angles which are 90° apart (Shu *et al.*, 1985b)

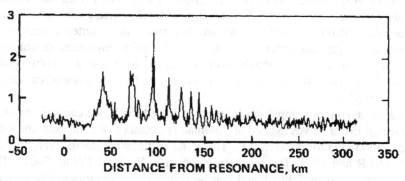

Figure 3. Non-linear density waves observed at Mimas 5:3 ILR in Saturn's A-ring. The ordinate is optical depth, a measurement of surface density.

Although the bar-driven density wave theory was motivated by galaxies with a major bar, the gasdynamic formulation is unlikely to solve the problem for them. This is because the stellar and the gas component are coupled in the galactic disks. Gasdynamics is not adequate for stars, especially near the Lindblad resonances. We must solve Jeans equation for the stellar disk and gasdynamic equations for the gas disk, separately but not independently, since the gravitational potential is determined jointly by stars and gas. Therefore, the resonance excitation mechanism can only give the qualitative understanding of the spiral-bar structure there.

On the other hand, the central regions (within 2–3 kpc) of disk galaxies provide an ideal laboratory for testing the theory. Due to the drastic difference in the dispersion velocities, stars and gas are almost completely decoupled there. Gas with sound speed typically of 10 km/s forms a disk, while the stars with dispersion speed of 100 km/s form a spherical bulge. The bar, as suggested in infrared observations, usually is composed of stars. The bar potential, whether of the size of the whole galaxy or of a kpc-size in the nucleus, can provide the periodic forcing to excite waves in the central gas disk. A fast bar will excite an OLR in the central region and impart positive angular momentum to the disk. For a slow bar, it will excite waves at ILR's and impart negative angular momentum (For more detail, see Polyachenko, 1994; Yuan and Kuo, 1997, 1999; Fridman and Khoruzhii, 2000). With viscous damping, angular momentum will deposit in a region covered by the waves. Disk material, gaining angular momentum, will move outward to form a ring of high density, and, losing angular momentum, will either form an oval ring around the center or to fall all the way to the center. Those rings would be the location to host starburst activities. Material going to the center will fuel the AGN, if the galaxy has a supermassive blackhole there. These issues are of great interest in modern astrophysics.

The first successful application of this approach was to interpret the expansion velocity of the 3-kpc Arm by assuming a rotating bar or oval distortion in the center of the Milky Way (Yuan, 1984; Yuan & Cheng, 1987). The central bar was later confirmed by observations (Blitz & Spergel, 1991). After that, a great many studies have been dedicated to the central bars and their resonantly excited spirals, almost all by numerical gasdyanamics (e.g., Sheth et al., 2000; Englmaier & Shlosman, 2000). This enthusiasm is obviously motivated by the high resolution observations by mm and sub-mm radio interferometry arrays and by new infrared cameras on ground or in space.

Using the asymptotic approach formulated in section 3, we can systematically study the non-linear waves excited at Lindblad resonances. We integrate the differential Equation (11) for different Ω_p. The spiral patterns are distinctively different at OLR, OILR and IILR. The results are shown in Figure 4 (Yuan, 2001). Their properties are listed in Table 1 and can be tested by observations. For instance, galaxies with open central spirals should have iso-velocity curves bent outward along spirals. Furthermore, to have ILR in the central region implies small Ω_p, or slow bars. Slow bars probably result from orbital trapping mechanism (Lynden-Bell, 1979; also in Fridman and Polyachenko, 1984). This mechanism requires less concentration of mass at center, which would give rise to a slow rising rotation curve. Thus, *slow rising rotation curve, open central spirals, and outward bent iso-velocity curves are correlated*. An example of this is found for NGC4321, shown in Figure 5. Similarly, *rapidly rising rotation curve, tightly wound spirals, outward bent iso-velocity curves along spirals are correlated*. So far, this theory has been applied to a number of galaxies with good agreement with observations

TABLE I

Characteristics of spirals and iso-velocity curve predicted for different types of the Lindblad resonances.

	bar rotation Ω_p	sense of winding	tightness of spirals	bending of iso-velocity curve
IILR	slow	leading	open	inward along spirals
OILR	slow	trailing	open	outward along spirals
OLR	fast	trailing	tight	inward along spirals

Figure 4. Spiral density waves excited by gravitational potential of a bar lying horizontally at, from left, IILR, OILR, and OLR. They are respectively open leading, open trailing and tight trailing

(Yuan & Kuo, 1998). Identifying a leading spiral pattern in the center of NGC157, corresponding to the IILR, was reported by Fridman and Khoruzhii *et al.* (2001).

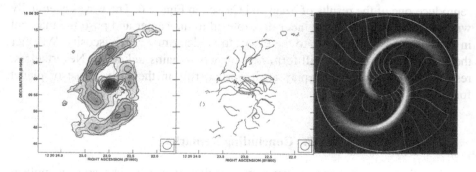

Figure 5. Comparison with observations of NGC4321 by Sakamoto *et al.* (1995). Left, grey-scale CO intensity; central, iso-velocity curves; right, theoretical results. Please note open trailing spirals with iso-velocity curves bent outward along spirals

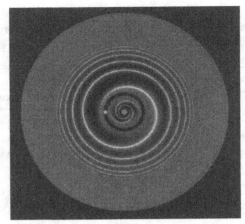

Figure 6. Proto-planet excites density waves in the proto-stellar disk. We have 2:1 OLR producing 1-arm spiral and 2:1 ILR producing 2-arm spirals. For Keplerian disks, the 2:1 OLR are referred to the case $(m + 1) : m$, with m=1, meaning an 1-arm spiral.

6. Spiral Waves in Proto-stellar Disks

With ever-improving observational techniques, proto-stars and proto-stellar disks are reportedly discovered. What would happen if a proto-planet is formed and embedded in the proto-stellar disk? It probably wipes out the particles (or planetesimals) next to it by either accreting them or scattering them. For disk regions reasonably away from the proto-planet, density waves would be excited by the proto-planet. These waves would cover parts of the disks and re-distribute the disk material by means of transporting angular momentum between the proto-planet and the disk. It will excite waves at both OLR and ILR. Lin and Papalozou (1986) and Yuan and Cassen (1992) addressed this problem. To demonstrate this process, we reproduce one of the results of Yuan and Cassen in Figure 6. The waves eventually will push the material in the inner disk toward to the center and push the material in the outer disk further out, to the extent to be determined by viscosity. Whether the accumulated material will form new planets remains unknown. Nevertheless, resonance excitation would play an important role in the early phase of planet formation.

7. Concluding Remarks

In this article, we examine the resonance excitation mechanism and its applications to Saturn's rings, galactic disks, and proto-stellar disk, through a non-linear asymptotic theory. In this formulation, the effects of viscosity and self-gravity can be explicitly included. Moreover, because we can reduce the two-dimensional disk problem to one-dimension in the WKBJ approximation, we calculate the problem with infinite resolution in space. All of these are extremely difficult to be achieved by numerical simulations, even for the one-dimensional problem. Furthermore, since the focus is on the Lindblad resonance, the analysis has been carried out conveniently by taking expansions at the resonance. Thus, the method used so far does not treat the global problem properly. To improve the method which is possible will cost us dearly. The situation worsens when the waves approach the center. Not only the curvature effect will come in, but also the wavelength will be comparable to the thickness of the disk. Unfortunately, following the waves to the center and tracking the mass flux going to the center hold the key to the understanding of fueling AGN's, one of the most important issues in modern astrophysics. How to overcome these difficulties yet remain in the framework of this asymptotic approach, will be a challenge for the time to come.

Numerical simulations, using advanced CFD techniques, can give us the answers to the disk evolution problem. However, to get reliable results is no simple matter either. The boundary conditions are well known to be difficult to handle even for the two-dimensional problems. Improper boundary conditions, how small they may be, are capable of corrupting good results. The problem is especially tricky near the center. Waves being reflected from the center are extremely difficult to get rid of. They often mislead us. After all we don't even know whether waves or portion of the waves are indeed reflected in nature. Another problem is the viscosity. In all schemes, viscosity enters implicitly. How much it has already in the scheme and how much more we should add to it without ruining the true answer is again a tricky business. Furthermore, we also have to include the self-gravitation component in the calculations.

In summary, resonance excitation is a powerful mechanism. It is responsible for the spiral structure in various disk systems in the Universe and it also determines their evolution. Using the asymptotic approach, we are able to bring out the fundamental physics of resonance excitation. Now we think we understand how a periodic perturber can excite spiral density waves at the Lindblad resonance and through these waves how the angular momentum can be exchanged between the disk and the perturber. Therefore it helps us resolve those important issues in astrophysics, such as how gaps can be cleared in planetary rings, how disk matter can be directed to the galactic center to fuel AGN's, and how a high density rings can be formed, which can host starburst activities in galaxies.

Notes

1. When viscosity is included, the coefficient for first term would be replaced by $(-a^2 + i\kappa\xi)$, where ξ is the kinematic viscosity.
2. For bending waves, we must change G by $-G$ and κ by μ.
3. This is called the angular quasi-momentum by Fridman and Gorkavyi (1999) and is the quantity we refer as angular momentum density of the waves in the context.

4. Fridman and Gorkavyi (1999) considered it and found it has contribution to the viscous stress of the disk.
5. We use transformation instead of reflection or refraction, because both wavelengths and phase change during the process. Neither reflection nor refraction can describe it properly.

Acknowledgements

I wish to thank Drs. C.C. Lin, Frank Shu and Bruce Elmegreen for helpful discussions. Alex Fridman has kindly read the early version of manuscript and made many valuable suggestions and comments, which help improve the quality of this paper. To him, I owe my thanks. Mr. Lupin Lin helped me prepare Figure 1 in this paper, for which I am grateful.

References

Bertin, G.: 2000, 'Dynamics of Galaxies', Cambridge U. Press, Cambridge: UK.
Bertin, G. and Lin, C. C.: 1996, 'Spiral Structure in Galaxies: A Density Wave Theory', MIT Press: Cambridge, MA.
Bertin, G., Lin, C. C., Lowe, S. A., and Thurstan, R. P.: 1989a, 'Modal approach to the morphology of spiral galaxies. I – Basic structure and astrophysical viability', *Astrophys. J.* **338**, 78.
Bertin, G., Lin, C. C., Lowe, S. A., and Thurstan, R. P.: 1989b, 'Modal Approach to the Morphology of Spiral Galaxies. II. Dynamical Mechanisms', *Astrophys. J.* **338**, 104.
Blitz, L. and Spergel, D. N.: 1991, 'Direct Evidence for a Bar at the Galactic center', *Astrophys. J.* **397**, 631.
Borderies, N., Goldreich, P., and Tremaine, S.: 1984, 'Unsolved problems in planetary ring dynamics', in *Planetary Rings*, Greenberg, R. and Brahic, A. (eds.), Tucson: University of Arizona Press, 713.
Buta, R., Crocker, D. A., and Elmegreen, B. G.: 1996, 'Barred Galaxies', IAU Colloquium 157, ASP: San Francisco, CA.
Combes, F., Mamon, G. A., Charmandaris, V.: 2000, 'Small Galaxy Groups', IAU Colloquium 174, ASP: San Francisco, CA.
Fridman, A. M. and Gorkavyi, N. N.: 1999, 'Physics of Planetary Rings', Springer-Verlag: Berlin, Heidelberg, New York.
Fridman, A. M. and Khoruzhii, O. V.: 2000, *Phys. Lett.* **A278**, 199.
Fridman, A. M. and Khoruzhii, O. V.: 2001, 'Restoring the full velocity field in the gaseous disk of the spiral galaxy NGC 157', *Astron. Astrophys.* **371**, 583.
Fridman, A. M. and Polyachenko, V. L.: 1984, 'Physics of Gravitating Systems', Springer, New York.
Englmaier, P. and Shlosman, I.: 2000, 'Density Waves inside the Inner Lindblad Resonance: Nuclear Spirals in Disk Galaxies', *Astrophys. J.* **528**, 677.
Goldreich, P. and Tremaine, S.: 1978a, 'The Velocity Dispersion in Saturn's Rings', *Icarus* **34**, 227.
Goldreich, P. and Tremaine, S.: 1978b, 'The Formation of the Cassini Division in Saturn's Rings', *Icarus* **34**, 240.
Goldreich, P. and Tremaine, S.: 1979, 'The Excitation of Density Waves at the Lindblad And Corotation Resonances by an External Potential', *Astrophys. J.* **233**, 857.
Goldreich, P. and Tremaine, S.: 1980, 'Disk-Satellite Interactions', *Astrophys. J.* **241**, 425.

Goldreich, P. and Tremaine, S.: 1982, 'The Dynamics of Planetary Rings', *Ann Rev Astron. & Astrophys.* **20**, 249.

Huntley, J. M., Sanders, R. H., Roberts, W. W.: 1978, 'Bar-Driven Spiral Waves in Disk Galaxies', *Astrophys. J.* **221**, 521.

Lin, C. C. and Shu, F. H.: 1964, 'On the Spiral Structure of Disk Galaxies', *Astrophys. J.* **140**, 646.

Lin, C. C. and Shu, F. H.: 1966, 'On the Spiral Structure of Disk Galaxies. II. Outline of a Theory of Density Waves', *Proc. Nat. Acad. Sci.* **55**, 229.

Lindblad, B.: 1963, *Stockholm Observ. Ann.* **22**, 3.

Lin, D. C. N. and Papalozou, J.: 1986, 'On the tidal interaction between proto-planets and the primordial solar nebula. II – Self-consistent nonlinear interaction', *Astrophys. J.* **307**, 395.

Lynden-Bell, D.: 1979, 'On a Mechanism that Structures Galaxies', *Monthly Notices of the Royal Astron. Soc.* **187**, 101.

Lynden-Bell, D. and Kalnajs, A. J.: 1972, 'On the Generating Mechanism of Spiral Structure', *Monthly Notices of the Royal Astron. Soc.* **157**, 1.

Mark, J. W.-K.: 1976, 'On Density Waves in Galaxies. III. Wave Amplification by Stimulated Emission', *Astrophys. J.* **205**, 363.

Miyama, S. M., Tomisaka, K., Hanawa, T.: 1999, 'Numerical Astrophysics', Kluwer Acad. Pub., Dordrecht.

Sakamoto, K., Okumura, S., Minezaki, T., Kobayashi, Y., and Wada, K.: 1995, 'Bar-Driven Gas Structure and Star Formation in the Center of M100', *Astron. J.* **110**, 2075.

Sandqvist, A., and Lindblad, P. O.: 1995, 'Barred Galaxies and Circumnuclear Activities', Springer-Verlag, Berlin.

Sheth, K., Regan, M. W., Vogel, S. N., and Teuben, P. J.: 2000, 'Molecular Gas, Dust, and Star Formation in the Barred Spiral NGC 5383', *Astrophys. J.* **532**, 221.

Shu, F. H.: 1984, 'In Planetary Rings', Greenberg, R. and Brahic, A. (eds.), U. Ariz. Press, Tucson, AR, 513.

Shlosman, I, Begelman, M. C., and Frank, J.: 1990, 'The fueling of active galactic nuclei', *Nature* **345**, 679.

Shu, F. H., Laughlin, G., Lizano, S.; Galli, D.: 2000, 'Singular Isothermal Disks. I. Linear Stability Analysis', *Astrophys. J.* **535**, 190.

Shu, F. H., Dones, L, Lissauer, J. J. Yuan, C, and Cuzzi, J. N.: 1985b, 'Nonlinear spiral density waves – Viscous damping', *Astrophys. J.* **299**, 542.

Shu, F. H. and Stewart, G. R.: 1985, 'The Collisional Dynamics of Particulate Disks', *Icarus* **62**, 360.

Shu, F. H., Yuan, C., Lissauer, J. J.: 1985a, 'Nonlinear Spiral Density Waves – An Inviscid Theory', *Astrophys. J.* **291**, 356.

Toomre, A.: 1969, 'Group Velocity of Spiral Waves in Galatic Disks', *Astrophys. J.* **158**, 899.

Toomre, A.: 1981, in *Structure and Dynamics of Normal Galaxies*, Fall, S.M. and Lynden-Bell, D. (ed.), *Cambridge U. Press, Cambridge, UK*.

Polyachenko, V.: 1994, 'Galactic Bars and Associated Structure', in *Physics of the gaseous and stellar disks of the galaxy, ASPC series, 66*, King, I. R. (ed.), p 103.

Yuan, C.: 1984, 'On the '3 Kiloparsec Arm' – Resonance Excitation of Linear and Nonlinear Waves by an Oval Distortion in the Central Region', *Astrophys. J.* **281**, 600.

Yuan, C.: 2001, 'Resonance Excitation in Disk Systems in Astrophysics', in *Stellar Dynamics: From Classic to Modern*, Ossipkov, L. P. and Nikiforov, I. I. (ed.), St. Petersburg University Press.

Yuan, C. and Cassen, P.: 1994, 'Resonantly driven nonlinear density waves in proto-stellar disk', *Astrophys. J.* **437**, 338.

Yuan, C. and Cheng, Y.: 1987, in *The Outer Galaxy*, Blitz, L. & Lockman, J. (ed.), Springer, Berlin, 144.

Yuan, C. and Cheng, Y.: 1989, 'Resonance Excitation of Spiral Density Waves in a Gaseous Disk. I – A Linear Theory, 1989, ApJ, 340, 216', *Astrophys. J.* **376**, 104.

Yuan, C. and Kuo, C. L.: 1997, 'Bar-driven Spiral Density Waves and Accretion in Gaseous Disks', *Astrophys. J.* **486**, 750.

Yuan, C., & Kuo, C. L.: 1998, 'Spiral Structure in the Central Disks of NGC 1068 and M100', *Astrophys. J.* **497**, 689.